T0146083

Producing Liquid Fuels from Coal

Prospects and Policy Issues

James T. Bartis, Frank Camm, David S. Ortiz

Prepared for the United States Air Force and the
National Energy Technology Laboratory of the
United States Department of Energy

Approved for public release; distribution unlimited

PROJECT AIR FORCE and
INFRASTRUCTURE, SAFETY, AND ENVIRONMENT

The research described in this report was sponsored by the United States Air Force under Contract FA7014-06-C-0001. Further information may be obtained from the Strategic Planning Division, Directorate of Plans, Hq USAF. It was also supported by the National Energy Technology Laboratory, United States Department of Energy, and was conducted under the auspices of the Environment, Energy, and Economic Development Program within RAND Infrastructure, Safety, and Environment.

Library of Congress Cataloging-in-Publication Data is available for this publication.

ISBN: 978-0-8330-4511-9

The RAND Corporation is a nonprofit research organization providing objective analysis and effective solutions that address the challenges facing the public and private sectors around the world. RAND's publications do not necessarily reflect the opinions of its research clients and sponsors.

RAND® is a registered trademark.

Cover photo courtesy of Peabody Energy Corporation.

© Copyright 2008 RAND Corporation

All rights reserved. No part of this book may be reproduced in any form by any electronic or mechanical means (including photocopying, recording, or information storage and retrieval) without permission in writing from RAND.

Published 2008 by the RAND Corporation
1776 Main Street, P.O. Box 2138, Santa Monica, CA 90407-2138
1200 South Hayes Street, Arlington, VA 22202-5050
4570 Fifth Avenue, Suite 600, Pittsburgh, PA 15213-2665
RAND URL: http://www.rand.org/
To order RAND documents or to obtain additional information, contact
Distribution Services: Telephone: (310) 451-7002;
Fax: (310) 451-6915; Email: order@rand.org

Preface

The increase in world oil prices since 2003 has prompted renewed interest in producing and using liquid fuels from unconventional resources, such as biomass, oil shale, and coal. This book focuses on issues and options associated with establishing a commercial coal-to-liquids (CTL) industry within the United States. The book describes the technical status, costs, and performance of methods that are available for producing liquids from coal; the key energy and environmental policy issues associated with CTL development; the impediments to early commercial experience; and the efficacy of alternative federal incentives in promoting early commercial experience. Because coal is not the only near-term option for meeting liquid-fuel needs, this book also briefly reviews the benefits and limitations of other approaches, including the development of oil shale resources, the further development of biomass resources, and increasing dependence on imported petroleum.

A companion document provides a detailed description of incentive packages that the federal government could offer to encourage private-sector investors to pursue early CTL production experience while reducing the probability of bad outcomes and limiting the costs that might be required to motivate those investors. (See Camm, Bartis, and Bushman, 2008.)

The research reported here was performed at the request of the U.S. Air Force and the U.S. Department of Energy. The Air Force sponsor was the Deputy Chief of Staff for Logistics, Installations and Mission Support, Headquarters, U.S. Air Force, in coordination with the Air Force Research Laboratory. The Department of Energy sponsor was the National Energy Technology Laboratory. Within RAND, it was conducted as a collaborative effort under the auspices of the Resource Management Program of RAND Project AIR FORCE and the RAND Environment, Energy, and Economic Development Program (EEED) within RAND Infrastructure, Safety, and Environment.

During the preparation of this book, the U.S. Congress and federal departments were considering alternative legislative proposals for promoting the development of unconventional fuels in the United States. This book is intended to inform those deliberations. It should also be useful to federal officials responsible for establishing civilian and defense research programs; to potential investors in early CTL production plants;

and to state, tribal, and local government decisionmakers who are considering the costs, risks, and benefits of early CTL production plants.

To promote broad access to this book, we have avoided detailed technology descriptions and have relegated supporting econometric analyses to the appendix and the companion volume.

This book builds on earlier RAND Corporation publications on natural resources and energy development in the United States. Most relevant are the following:

- *Oil Shale Development in the United States: Prospects and Policy Issues* (Bartis et al., 2005)
- *Understanding Cost Growth and Performance Shortfalls in Pioneer Process Plants* (Merrow, Phillips, and Myers, 1981)
- *New Forces at Work in Mining: Industry Views of Critical Technologies* (Peterson, LaTourrette, and Bartis, 2001).

RAND Project AIR FORCE

RAND Project AIR FORCE (PAF), a division of the RAND Corporation, is the U.S. Air Force's federally funded research and development center for studies and analyses. PAF provides the Air Force with independent analyses of policy alternatives affecting the development, employment, combat readiness, and support of current and future aerospace forces. Research is conducted in four programs: Force Modernization and Employment; Manpower, Personnel, and Training; Resource Management; and Strategy and Doctrine.

Additional information about PAF is available on our Web site: http://www.rand.org/paf/

The RAND Environment, Energy, and Economic Development Program

The mission of RAND Infrastructure, Safety, and Environment is to improve the development, operation, use, and protection of society's essential physical assets and natural resources and to enhance the related social assets of safety and security of individuals in transit and in their workplaces and communities. The EEED research portfolio addresses environmental quality and regulation, energy resources and systems, water resources and systems, climate, natural hazards and disasters, and economic development—both domestically and internationally. EEED research is conducted for government, foundations, and the private sector.

Information about EEED is available online (http://www.rand.org/ise/environ).

Questions or comments about this book should be sent to the project leader, James T. Bartis (James_Bartis@rand.org).

Contents

Figures

Tables

Summary

During 2007 and 2008, world petroleum prices reached record highs, even after adjusting for inflation. Concerns about current and potentially higher future petroleum costs for imported oil have renewed interest in finding ways to use unconventional fossil-based energy resources to displace petroleum-derived gasoline and diesel fuels. If successful, this course of action would lower prices and reduce transfers of wealth from U.S. oil consumers to foreign oil producers, resulting in economic gains and potential national-security benefits.

Oil shale, tar sands, biomass, and coal can all be used to produce liquid fuels. Of these, coal appears to show the greatest promise, considering both production potential and commercial readiness. It is the world's most abundant fossil fuel. Global, proven recoverable reserves are estimated at one trillion tons (World Energy Council, 2004), which represent nearly three times the energy of the proven reserves of petroleum.

The technology for converting coal to liquid fuels already exists. Commercial coal-to-liquids (CTL) production has been under way in South Africa since the 1950s. Moreover, CTL production appears to be economically feasible at crude oil prices well below the prices seen in 2007 and 2008. However, without effective measures to manage greenhouse-gas emissions, the production and use of coal-derived liquids to displace petroleum-derived transportation fuels could roughly double the rate at which carbon dioxide is released into the atmosphere. In the absence of an effective national program to limit greenhouse-gas emissions, it is unclear whether the federal government would support the development of a CTL industry capable of producing millions of barrels per day (bpd) of liquid fuels.

Research Goals and Methodology

This study analyzed the costs, benefits, and risks of developing a U.S. CTL industry that is capable of producing liquid fuels on a strategically significant scale. Our research approach consisted of the following basic steps:

- To understand commercial development prospects, we examined what is known and not known regarding the economic and technical viability and the environmental performance of commercial-scale CTL production plants.
- To quantify benefits and understand how the large-scale introduction of unconventional fuel sources might affect both the world price of oil and the well-being of oil consumers and producers, we developed a model of the global oil market designed to allow us to compare policy alternatives in the face of inherent uncertainties about how various aspects of the market might behave in the future.
- To investigate how integrated packages of public policy instruments could encourage investment in unconventional-fuel production plants, we reviewed fundamental aspects of contract design and developed a financial model to determine how those incentive packages might affect (1) the rate of return to investors and (2) the net present value of cash flows between such plants and the government.

Finally, our study consistently took into account two overarching policy goals: reducing dependence on imported oil and decreasing greenhouse-gas emissions.

Principal Findings

U.S. Coal Resources Can Support a Domestic Coal-to-Liquids Industry Far into the Future

The United States leads the world with recoverable coal reserves estimated at approximately 270 billion tons. These reserves are broadly distributed, with at least 16 states having sufficient reserves to support commercial CTL production plants (see pp. 9–12). In 2006, the United States mined a record 1.16 billion tons of coal, nearly all of which was used to produce electric power. Dedicating only 15 percent of recoverable coal reserves to CTL production would yield roughly 100 billion barrels of liquid transportation fuels, enough to sustain three million bpd of CTL production for more than 90 years (see pp. 12–13).

Technology for Producing Coal-to-Liquids Fuels Has Advanced in Recent Years

In the United States, interest in CTL fuels has concentrated on two production approaches that begin with coal gasification: the Fischer-Tropsch (FT) and methanol-to-gasoline (MTG) liquefaction methods. The FT method was invented in Germany during the 1920s and is in commercial practice in South Africa. The Mobil Research and Development Corporation invented the MTG approach in the early 1970s. Both approaches involve preparing and feeding coal to a pressurized gasifier to produce *synthesis gas*—the important constituents of which are hydrogen and carbon monoxide. After deep cleaning, processing, and removal of carbon dioxide, the synthesis gas is sent to a catalytic reactor, where it is converted to liquid hydrocarbons. The principal products of an FT CTL plant are exceptionally high-quality diesel and jet fuels that can be sent directly to local fuel distributors (see pp. 20–22). In an MTG CTL plant, the synthesis gas is first converted to methanol. The methanol is then converted to a mix of hydrocarbons that are very similar to those found in raw gasoline. Between 90 and 100 percent of the final liquid yield of an MTG CTL plant is a zero-sulfur automotive gasoline that can be distributed directly from the plant. (See pp. 25–26.)

A favorable attribute of both approaches is that the synthesis gas can be produced from a variety of feeds, including natural gas, biomass, and coal. Although no FT CTL plants have been built in more than 20 years, the FT approach has advanced through the recent and ongoing construction of large commercial plants designed to produce liquids from natural gas that cannot be pipelined to nearby markets (see p. 19). Although no commercial MTG CTL plant has ever been built, we judge the process as ready for initial commercial operations, based on ten years of large-scale operating experience, starting in 1985, when the process was commercially applied to produce gasoline from natural-gas deposits in New Zealand (see pp. 24–25).

Technology for Controlling Carbon Dioxide Emissions Is Advancing

If the entire fuel cycle is taken into account—i.e., oil well or coal mine through production to end use—we estimate that greenhouse-gas emissions from a CTL plant would be about twice those associated with fuels produced from conventional crude oils. Slightly higher values would result from less efficient CTL plants or by comparing with light crude oils. And slightly lower values would result from more energy-efficient CTL plant designs or by comparing with the heavier crude oils that are taking an increasing role in worldwide oil production. Technological advances aimed at significantly improving the energy efficiency and costs of CTL production might be able to reduce plant-site greenhouse-gas emissions by one-fifth—not enough to match those of conventional petroleum (see pp. 31–32). To avoid conflict with growing national and international priorities to reduce global greenhouse-gas emissions, the large-scale development of a CTL industry requires management of plant-site carbon dioxide emissions.

Capturing the carbon dioxide that would be otherwise emitted from a CTL plant is straightforward and relatively inexpensive. CTL plants already remove carbon dioxide from the synthesis gas, so capture simply involves dehydrating and compressing the carbon dioxide so that it is ready for pipeline transport. If 90 percent of plant-site emissions were to be fully captured and then stored, the production and use of fuels produced in early CTL plants should not cause any significant increase or decrease in greenhouse-gas emissions as compared to fuels derived from conventional light crude oils. For nearly full capture of plant-site carbon dioxide emissions, we estimate that product costs would increase by less than $5.00 per barrel. (See pp. 32–33.)

There are two principal methods for disposing of the captured carbon dioxide. The first is to use the captured carbon dioxide to enhance oil recovery in partially depleted oil reservoirs using a well-known method called carbon dioxide flooding. The advantage of this method is that at least two barrels of additional conventional petroleum will be produced for each barrel of CTL fuel. Moreover, CTL plant operators might be able to sell their captured carbon dioxide at a profit above their costs of capture and transport. This enhanced oil recovery method is limited to the first 0.5 million bpd to one million bpd of CTL production capacity built within a few hundred miles of appropriate oil reservoirs. A pioneer field test and demonstration of carbon dioxide sequestration through enhanced oil recovery has been under way since 2000 at the Weyburn oil field in Saskatchewan. (See pp. 34–36.)

The second method is to sequester carbon dioxide in various types of geologic formations. The latter approach is broadly viewed as the critical technology that will allow continued coal use for power generation while reducing greenhouse-gas emissions. Two major demonstrations of carbon dioxide sequestration in geological formations are under way outside the United States. Results to date have been promising (see p. 36). However, the development of a commercial sequestration capability within the United States requires addressing important knowledge gaps associated with site selection and preparation, predicting long-term retention, and monitoring and modeling the fate of the sequestered carbon dioxide. There are also important legal and public acceptance issues that must be addressed. Toward this end, U.S. Department of Energy plans to conduct at least eight moderate- to large-scale demonstrations over the next five years. (See pp. 74–75.)

A Combination of Coal and Biomass to Produce Liquid Fuels May Be a Preferred Solution

Biomass can be converted to a synthesis gas that FT reactors can use to produce fuels identical to those derived from coal or natural gas. The biomass-to-liquids (BTL) approach results in low total-fuel-cycle release of greenhouse gases because the emissions at the plant are balanced by the carbon dioxide absorbed from the atmosphere during the growth cycles of the biomass crops.

A promising direction for alternative-fuel production would be an integrated FT or MTG plant designed to accept both biomass and coal. A coal- and biomass-to-liquids (CBTL) approach can ameliorate problems created by the use of biomass alone—i.e., the logistics of biomass delivery that limit production levels and the annual climate variations that can cause major fluctuations in the quantity of biomass available to a BTL-only plant. A CBTL plant can be substantially larger than a BTL plant, and its large-scale economies would enable it to operate at a significantly lower cost. The marginal benefits of adding a coal feedstock to a biomass feedstock may more than offset the marginal costs associated with sequestering the increased carbon dioxide emissions that result. (See pp. 37–38.)

Given information that is currently available and considering the entire fuel cycle, we conclude that CBTL fuels can be produced *and used* at greenhouse-gas emission levels that are well below those associated with the production and use of conventional petroleum fuels. For example, with 90-percent sequestration of plant-site emissions, we estimate that a 55/45 coal/dry biomass mix (based on energy input) will result in CBTL fuel production with zero net greenhouse-gas emissions considering the full fuel cycle from coal mining and biomass cultivation to end use. Likewise, a 75/25 coal/dry biomass mix would yield roughly a 55- to 65-percent reduction in greenhouse-gas emissions, as compared to conventional petroleum fuels. (See pp. 39–40.)

Developing a Coal-to-Liquids Industry in the United States Will Be Expensive, but Significant Production Is Possible by 2030

CTL plants are capital intensive. For moderate to large CTL plants, we estimate capital investment costs of $100,000 to $125,000 (in January 2007 dollars) per barrel of product. Considering operating and coal costs, we estimate that, for CTL fuels to be competitive, the selling price for crude oil (using a West Texas Intermediate benchmark) must be between $55 and $65 per barrel. These prices include the costs of capturing about 90 percent of carbon dioxide emissions but do not assume any income or outlays associated with sequestering that carbon dioxide. Our cost estimates are highly uncertain, since they are based on low-definition engineering designs. Also, our estimates apply only to the first generation of CTL plants built in the United States. We expect the cost of building and operating new plants to drop significantly once early commercial plants begin production and experience-based learning is under way. (See pp. 42–45.)

Considering the importance of experience-based learning, the need to avoid cost-factor escalation, and the time required to bring carbon capture and sequestration to full commercial viability, we estimate that, by 2020, the production level of CTL fuels can be no more than 500,000 bpd. Post-2020 capacity buildup could be rapid, with U.S.-based CTL production potentially in the range of three million bpd by 2030. (See pp. 46–48)

Coal-to-Liquids Development Offers Strategic National Benefits

The United States now consumes about 20 million barrels of liquid fuels per day. This level of use is projected to rise slightly over the next 25 years. If a domestic CTL industry is developed and operates on a profitable basis, the United States would benefit from the economic profits generated by that industry. CTL production would benefit oil consumers by reducing the world price of oil, and this reduction in world oil prices would yield national security benefits. Having a domestic CTL industry in place would also increase the resiliency of the petroleum supply chain in the United States and provide enhanced employment opportunities, especially in states holding large reserves of coal. To examine these benefits, we assumed a hypothetical domestic CTL production rate of three million bpd by 2030.

Economic Profits. If a large CTL industry develops by 2030, we anticipate that post-production learning will result in significantly lower CTL production costs. At world crude oil prices of between $60 and $100 per barrel (2007 dollars), direct economic profits of between $20 billion and $70 billion per year are likely. Through various taxes, a portion of these profits, between $7 billion and $25 billion per year, would go to federal, state, and local governments and thereby broadly benefit the public. (See p. 60.)

Reduced World Oil Prices. Lower world oil prices will likely be the result of any increase in liquid-fuel production, either domestically or abroad, from unconventional resources. Based on examining a broad range of potential responses by the Organization of the Petroleum Exporting Countries (OPEC), we anticipate that world oil prices will drop by between 0.6 and 1.6 percent for each million barrels of unconventional-fuel production that would not otherwise be on the market. Further, this price decrease should be close to linear for unconventional-fuel additions of up to ten million bpd. Unconventional-fuel additions in this range are possible, but only by considering potential 2030 production levels from domestic oil shale and biofuel resources as well as both domestic and international production of coal-derived liquid fuels. Looking only at coal-derived liquids, it is possible that total world production could reach about six million bpd by 2030. (See p. 62.)

By reducing oil prices, consumer and business users of oil in the United States (and elsewhere) would benefit. From a national perspective, reduced profits to domestic petroleum producers would offset a portion of these benefits. Considering both oil users and producers, we estimate a net national benefit at between $2 billion and $8 billion per year for each million barrels per day of unconventional-fuel production (see pp. 63–65). Or equivalently, by lowering world oil prices, each barrel of CTL benefits the overall economy by between $6 and $24. The estimate of these benefits reflects our judgment that long-term oil prices will range between $60 and $100 per barrel with a range of market responses to the added supplies of liquid fuels. These benefits accrue to the nation as a whole, as opposed to the individual firms investing in CTL production. These analytic results support our finding that, to counter efforts of certain foreign oil suppliers to control prices by restraining production, the United States should be willing to spend $6 to $24 per barrel more than market prices for substitutes that reduce oil demand. (See pp. 65–66.)

National Security Benefits. The national security benefits of having a domestic CTL industry in place flow primarily from the anticipated reduction in world oil prices and thereby a reduction in revenues to oil-exporting countries. To the extent that this reduction in prices and revenues helps to limit behavior counter to U.S. national interests, there would be a benefit beyond the economic gain in reduced oil prices just noted. However, a three million bpd domestic industry would yield between a 3- and 8-percent reduction in the revenues of oil exporters. This small change in revenue would unlikely change the political dynamics in oil-producing nations unfriendly to the United States. With regard to enhancing national security, the principal contribution of CTL production would be its role in a portfolio of measures to increase liquid-fuel supplies and reduce oil demand. For example, global unconventional-fuel production of ten million bpd by 2030 could reduce OPEC annual revenues by up to a few hundred billion dollars. (See pp. 66–67.)

Environmental Impacts of a Large-Scale Coal-to-Liquids Industry Will Need to Be Addressed

Under current federal and state environmental, reclamation, and safety laws and regulations, the land, air, water, and ecological impacts of coal mining are mitigated to varying degrees. However, residual impacts of mining activities can still adversely change the landscape, the local ecology, and water quality. CTL development at a scale of three million bpd by 2030 would require about 550 million tons of coal production annually. Depending on whether and how greenhouse-gas emissions are controlled during this period, the net change in coal production between now and 2030 resulting from a gradual buildup of demand from a CTL industry could range from minimal up to an increase of about 50 percent above current levels. If large-scale development of a CTL industry is accompanied by a significant net increase in coal production or a significant change in extraction technologies, a review of the legislation and regulations governing mine safety, environmental protection, and reclamation may be appropriate. Such a review would assess the potential environmental and safety impacts of increased mining activity and evaluate options for reducing such impacts. More immediately, there is a clear need for research directed at mitigating the known and anticipated environmental impacts and reducing the work hazards associated with coal mining. (See pp. 78–79.)

Because of advances in environmental control technologies, CTL plant operations should not pose significant threats to air and water quality. There will be some locations where CTL development will be limited or prohibited, but, given the geographic diversity of the domestic coal resource base, large-scale development is unlikely to be impaired by a lack of suitable plant sites. (See pp. 76–78.)

It is difficult to predict how future, more technically mature CTL plants would manage water supply and consumption, especially in arid regions of Montana and Wyoming that hold enormous coal resources. Although design options are available to reduce water use in CTL plants, water consumption may be a limiting factor in locating multiple CTL plants in arid areas. (See pp. 79–81.)

Uncertainties Are Impeding Private Investment

Although numerous private firms have expressed considerable interest in CTL development in the United States, actual investment levels appear to be very limited. Discussions with proponents of CTL development indicate that three major uncertainties are impeding private investments:

- uncertainty about CTL production costs
- uncertainty regarding how and whether to control greenhouse-gas emissions
- uncertainty regarding the future course of world crude oil prices.

Of these three factors, the greatest impediment appears to be the uncertainty regarding future world oil prices. If investors would be confident that average long-term crude prices would remain consistently above $100 per barrel, no government policy would be required to support the emergence of a successful commercial CTL industry. But with the possibility that oil prices could fall significantly in the near to medium term, the financial risk surrounding initial CTL investments is appreciable. Given the extremely large capital investment required for even a moderate-size CTL plant, very few firms have the financial resources to take on this risk. (See p. 82.)

To Spur Early Coal-to-Liquids Production Experience, Government Incentives Should Target Prevailing Uncertainties

The firms most capable of overseeing the design, construction, and operation of CTL plants are the major petrochemical companies, which have the technical capabilities and the financial and management experience necessary for investing in multibillion-dollar megaprojects. They are also best suited to exploit the learning that would accompany early production experience. Yet none has announced interest in building first-of-a-kind CTL plants in the United States. (See p. 81.)

How can the federal government encourage the early participation of these and other capable companies in the CTL enterprise? The answer lies in the creation of incentive packages that cost-effectively transfer a portion of investment risks to the federal government.

We found that a balanced package of a price floor, an investment incentive, and an income-sharing agreement is well suited to do this. The investment incentive, such as a tax credit, is a cost-effective way to raise the private, after-tax internal rate of return in any future. A price floor provides protection in futures in which oil prices are especially low. And an income-sharing agreement compensates the government for its costs and risk assumption by providing payments to the government in futures in which oil prices turn out to be high (see pp. 92–96). Because the most desirable form of a balanced package depends on expectations about project costs, the government should wait to finalize its design until it has the best information on project costs that is available without actually initiating the project. Specifically, an incentive agreement should not be finalized until both government and investors have the benefit of improved project-cost and performance information that would be provided at the completion of a front-end engineering design. (See pp. 96–97.)

Loan guarantees can strongly encourage private investment. However, they encourage investors to pursue early CTL production experience only by shifting real default risk from private lenders to the government. By their very nature, the more powerful their effect on private participation, the higher the expected cost of these loan guarantees to the government. In addition, loan guarantees encourage private investors to seek higher debt shares that increase the risk of default and thus increase the government's expected cost for providing the guarantee. The government should take great care in employing loan guarantees to promote early CTL production experience. It should fully recognize both the costs that such guarantees could impose on taxpayers and the extent to which government oversight of guaranteed loans can be effective in limiting those costs. (See pp. 98–100.)

Overall Prospects

The prospects for developing an economically viable CTL industry in the United States are promising, although important uncertainties exist. Both FT and MTG CTL technologies are ready for initial commercial applications in the United States; production costs appear competitive at world oil prices well below current levels; and proven coal reserves in the United States are adequate to support a large CTL industry operating over the next 100 years.

Opportunities to control greenhouse-gas emissions from CTL plants are currently limited to enhanced oil recovery. But the prospects for successful development of large-scale geologic sequestration are promising, as is the development of technology that would allow the combined use of coal and biomass in production plants based on either the FT or MTG approaches. Within a few years, CTL plants could begin to alleviate growing global dependence on price-controlled conventional petroleum at greenhouse-gas emission levels comparable to those associated with conventional-petroleum products. Within a few more years, we anticipate that approaches would be available that allow the combined use of coal and biomass to produce liquid fuels so that total-fuel-cycle greenhouse-gas emission levels are significantly below those associated with conventional petroleum. (See pp. 46–48.) Most importantly, the low cost of capturing carbon dioxide at CTL plants implies that any measure that will induce reductions in greenhouse-gas emissions from coal-fired power plants will also be more than adequate to promote deep removal at CTL or CBTL plants (see p. 74).

Key Recommendations

With regard to the development of coal-derived liquids or other unconventional-fuel sources, the government could place itself anywhere along a continuum of policy positions. At one extreme is the hands-off position, which favors the free operation of the market and private decisionmaking unfettered by government interference. Support would be available for long-term research and development directed at significantly improving the economic and environmental performance of CTL production but not for near-term technology development or demonstration activities. (See pp. 106–107.)

At the other extreme, the government could choose specific alternatives to conventional oil production today and initiate large-scale federal support of these alternatives to ensure successful development of a new industry that can displace conventional oil production in the world market over the long term. (See pp. 107–109.)

Our research supports a policy strategy that falls between these extremes. This *insurance-policy* approach recognizes prevailing uncertainties and emphasizes future capabilities. The five elements of the insurance strategy are as follows:

- Cost-share a few site-specific front-end engineering design studies of CTL and dual-feedstock production plants to establish costs, risks, potential economic performance, and environmental impacts. (See p. 109.)
- Use federal incentives to ensure early commercial production experience with a limited number of first-of-a-kind CTL or dual-feedstock plants to establish performance and provide a foundation for post-production learning. (See p. 110.)
- Conduct multiple large-scale, long-term demonstrations of the sequestration of carbon dioxide generated at electricity-generation or CTL production plants (or both) at a scale and duration beyond that currently planned for in the U.S. Department of Energy's Regional Carbon Sequestration Partnerships. (See p. 111.)
- Undertake the research, development, and testing required to establish the technical viability of using a combination of biomass and coal for liquid production. (See p. 112.)
- Broaden and expand the federal portfolio directed at long-term, high-payoff research relevant to transportation fuel production. (See p. 112.)

The principal value of federal efforts to implement an insurance strategy is to accelerate CTL commercial development above what it would otherwise be. A five-year acceleration of development of a strategically significant CTL industry in the United States could result in national economic benefits with a present value of about $100 billion. (See p. 117.)

Air Force Options for Coal-to-Liquids Industrial Development

Should the Air Force choose to play an active role in promoting the development of a domestic CTL industry, it should do so recognizing that the primary potential benefits of success would accrue more to the nation as a whole than to the Air Force as an institution. (See p. 113.)

The U.S. Air Force's 2016 goal of being prepared to acquire alternative fuel blends to meet 50 percent of its domestic aviation fuel requirements is consistent with an overall federal insurance-policy strategy. The amount of FT CTL capacity required to meet the potential fuel purchases associated with the U.S. Air Force goal (50,000–80,000 bpd) falls within the overall production requirements of an insurance strategy, namely, obtaining early production experience from a limited number of CTL plants. (See p. 106.)

The U.S. Air Force might consider using fuel purchase contracts to promote early CTL production experience. To be more cost-effective, such contracts should be part of a broader federal package of investment incentives, such as investment tax credits, accelerated depreciation, and loan guarantees. These additional instruments could allow lower price floors and lessen the probability of out-year government purchases at above-market prices. (See pp. 92–93.)

Another option that the U.S. Air Force and the U.S. Department of Defense (DoD) might consider is to use DoD's contracting authority to establish a guaranteed or fixed price over a significant portion of the operating life of a CTL plant. Such agreements are rarely observed in contracts between private parties. Our findings indicate that a long-term price guarantee should be avoided because it is among the least cost-effective approaches available to the federal government. (See p. 114.)

Currently, DoD contracts are limited by law (10 USC 2306b) to a duration of no more than five years, with options for an additional five years, and a total amount of less than $500 million, unless specifically authorized otherwise by Congress. As such, DoD's ability to provide incentives for private investments in early CTL plants is severely limited. New legislative authority is needed if DoD and the U.S. Air Force wish to overcome the limitations imposed on contract duration and size. (See p. 114.)

Acknowledgments

The authors benefited greatly from the technical expertise, business insights, and policy perspectives that were generously provided by the many firms, organizations, and people we contacted or met during the course of the research that led to publishing this book.

We extend our thanks to the technical experts who provided information critical to our work. Throughout our research, we have been able to access technical expertise available at or through the National Energy Technology Laboratory, including Daniel Cicero, Terry Ackman, Gary Stiegel, and Charles Drummond from the National Energy Technology Laboratory; John Winslow and Edward Schmetz of Leonardo Technologies; and David Gray, Charles White, and Glen C. Tomlinson of Noblis. We also thank James A. Luppens from the U.S. Geological Survey; Lowell Miller from U.S. Department of Energy headquarters; James (Tim) Edwards from the Air Force Research Laboratory; Patsy (Pat) Muzzell of the Army Tank-Automotive Research, Development and Engineering Center; Carl Mazza from the U.S. Environmental Protection Agency; Thompson M. Sloane from General Motors; Richard A. Bajura of West Virginia University; and Malcolm Weiss of MIT.

Among representatives of firms we contacted, we owe special thanks to Hunt Ramsbottom and Richard O. Sheppard of Rentech; Rosemarie Forsythe and Samuel A. Tabak of Exxon Mobil; Donald Paul of Chevron; Robert C. Kelly of DKRW Advanced Fuels; Jim Rosborough of Dow Chemical; and John W. Rich Jr. of WMPI.

During the course of our research, we had the opportunity to discuss our approach and early findings with congressional staff members and federal and state officials. The questions and issues they raised provided important focus to our efforts. Our thanks go to Michelle Dallafior of the U.S. House of Representatives Committee on Science and Technology, Governor Joe Manchin III of West Virginia, Governor Dave Freudenthal of Wyoming, and the Honorable Jody Richards of the Kentucky House of Representatives.

As part of our research, we conducted a small workshop on the status of carbon capture and sequestration technology. We thank the attendees for their participation, with special acknowledgment of the presenters: Michael L. Godec (Advanced Resources International), Larry R. Myer (Lawrence Berkeley National Laboratory),

Sean I. Plasynski (National Energy Technology Laboratory), and Pamela Tomski (EnTech Strategies).

Kathryn Zyla of the World Resources Institute kindly organized an opportunity for us to present our preliminary findings to members of the Washington, D.C., environmental research and advocacy community. We especially thank Elizabeth Martin Perera and David Hawkins of the Natural Resources Defense Council and Ben Schreiber of Environment America for their comments.

Our research efforts have been greatly enhanced by the support and encouragement provided by senior officials from the U.S. Air Force—in particular, Michael Aimone, William E. Harrison, and Paul Bollinger.

We also acknowledge the important contributions from J. Allen Wampler and General Richard L. Lawson (U.S. Air Force, ret.), who collaborated with us throughout the study leading to this book, providing important assistance, guidance, and a critical eye to our findings. We have also greatly benefited from the formal review of our manuscript by Keith Crane and Michael Toman of RAND and independent reviewers James M. Ekmann and Hillard Huntington.

Abbreviations

ANL	Argonne National Laboratory
BLM	Bureau of Land Management
bpd	barrels per day
BTL	biomass to liquids
Btu	British thermal unit
CBTL	coal and biomass to liquids
CDE	carbon dioxide equivalent
CTL	coal to liquids
DoD	U.S. Department of Defense
DVE	diesel value equivalent
EDS	Exxon donor solvent
EEED	RAND Environment, Energy, and Economic Development Program
EIA	Energy Information Administration
EPC	engineering, procurement, and construction
FT	Fischer-Tropsch
GREET	Greenhouse Gases, Regulated Emissions, and Energy Use in Transportation
GTL	gas to liquids
HHV	higher heating value
IGCC	integrated gasification combined cycle
ISE	RAND Infrastructure, Safety, and Environment

kWh   kilowatt-hour

LHV   lower heating value

LNG   liquefied natural gas

LPG   liquefied petroleum gas

MACRS   Modified Accelerated Cost Recovery System

MIT   Massachusetts Institute of Technology

MTG   methanol to gasoline

MW   megawatt

MWh   megawatt-hour

OECD   Organisation for Economic Co-Operation and Development

OMB   Office of Management and Budget, Executive Office of the President

OPEC   Organization of the Petroleum Exporting Countries

ppm   parts per million

psi   pounds per square inch

R&D   research and development

RD&D   research, development, and demonstration

SRC   solvent-refined coal process

ULSD   ultralow-sulfur diesel

USDA   U.S. Department of Agriculture

USGS   U.S. Geological Survey

CHAPTER ONE

Introduction

Rising petroleum prices have once again prompted interest in using coal to manufacture liquid fuels that can displace petroleum-derived gasoline and diesel fuels. Coal is abundant in the United States and throughout the world. Coal-to-liquids (CTL) technology is ready for initial commercial applications in the United States, and production appears to be economically feasible at recent crude oil prices, which during 2008 were well over $100 per barrel for West Texas Intermediate crude oil. These considerations suggest that using coal to produce liquid fuels can stanch the large transfers of wealth from oil consumers to oil producers, thus providing significant benefits to U.S. consumers and potentially enhancing U.S. national security. But there is also opposition to the concept of transforming coal to liquids. Without measures to manage carbon dioxide emissions, the use of coal-derived liquids to displace petroleum fuels for transportation will roughly double greenhouse-gas emissions. In this view, promoting CTL development is not compatible with the need to reduce emissions of the principal greenhouse gases that are widely believed to accelerate global climate change.

The research reported here investigated the costs and benefits of developing an industry within the United States that is capable of producing coal-derived liquid fuels on a strategically significant scale. By *strategically significant*, we mean production of several million barrels per day (bpd), so that CTL would meet an appreciable fraction of the roughly 20 million bpd of liquid fuels that are currently consumed in the United States. Early in the course of our research, we realized that we needed a better understanding of how a CTL industry would influence the world oil market. To quantify benefits, we modeled how additional supplies of unconventional fuels might affect oil prices, reduce U.S. consumer expenditures, and reduce Organization of the Petroleum Exporting Countries (OPEC) export revenues. To understand costs, we examined what is known and not known regarding the economic and technical viability and the environmental performance of commercial-scale CTL production plants. Throughout, our emphasis has been on two overarching policy objectives: substituting alternative fuels for petroleum and decreasing greenhouse-gas emissions.

An important goal of our research is to provide the U.S. Department of Energy and the U.S. Air Force with an analytical framework for deciding whether government promotion of a CTL industry is in the national interest and, if so, how best to undertake that promotion. To further explore this issue, we examined impediments to private-sector investment in CTL, and we conducted a quantitative analysis of alternative financial incentive packages by determining how various incentives motivate private-sector investment and pose costs and risks to the government.

During the course of this research, we transmitted preliminary findings through discussions with and progress reports to our sponsors and to the broader community interested in alternative fuels. The ensuing dialogue provided the study team with important insights and helped us focus this book on the critical issues.

About This Book

Chapter Two briefly reviews the U.S. and global coal resource base. A key issue is the extent to which coal resources can support production of transportation fuels and still fulfill coal's traditional role of fueling electric power generation. In Chapter Two, our focus is on geology—namely, the size and geographic distribution of coal resources. Later, in Chapter Six, we discuss environmental issues that may limit access to these resources.

Although CTL technology is in commercial use in South Africa, serious issues remain regarding the risks, costs, and performance of plants that might be built in the United States. Chapter Three addresses these issues as part of a review of the technical approaches for producing liquids from coal. There we also examine the viability of technical options for reducing greenhouse-gas emissions associated with the production or use of coal-derived liquids. Chapter Three also contains a timeline for the initial commercial operation and industrial buildup to establish a CTL industry in the United States that would be capable of supplying a few million bpd of fuel.

To understand the costs and benefits of governmental policies that might promote or deter the development of a CTL industry, it is necessary to understand the broader range of options for reducing dependence on conventional petroleum. These options are examined in Chapter Four. The primary emphasis is on other alternative fuels, such as oil shale and biomass-derived fuels.

If changes in public policy could increase the production of nonconventional fuel substitutes for petroleum, what would happen to the world market price of petroleum? And if such increases in production reduced the world market petroleum price, how would that price reduction affect the well-being of petroleum consumers in the United States or the members of OPEC? To address these questions, we developed a simple, transparent model of the global petroleum market that allows us to incorporate a range of behavioral responses. The results of this model are presented and discussed in Chapter Five as part of a broader review of the possible benefits of developing a CTL industry.

Beyond benefits, there are significant environmental issues associated with the development of a large CTL industry. Chapter Six addresses environmental issues, including greenhouse-gas emissions, adverse impacts of increased coal production, air and water quality impacts from production plants, and water demand.

Numerous proposals have been put forth for subsidizing or mandating the production of alternative fuels, including CTL. Chapter Seven describes a way to analyze and assess policy alternatives appropriate for promoting private-sector investments and gaining operating experience in early CTL plants. It first asks how different types of incentives would affect the behavior of a private investor and government agency working together to achieve this objective. It then explores ways in which such incentives, and the policies required to implement them, would affect the financial interests of both the investor and the government in a range of potential futures. This analysis yields recommendations about how to design an integrated package of policies to promote early production of alternative fuels.[1]

Chapter Eight synthesizes the principal findings of our study. It also presents a framework for decisionmaking along with suggestions regarding how that framework might be implemented. Finally, we recommend ways of moving forward with CTL that take into account potential benefits as well as risks and uncertainties.

---

[1]   The topics addressed in Chapter Seven are discussed in much greater detail in the companion report, Camm, Bartis, and Bushman (2008).

CHAPTER TWO

The Coal Resource Base

Of the major fossil fuels, coal is the most abundant. Global, proven recoverable reserves are estimated at one trillion tons (World Energy Council, 2004), nearly triple the energy of the world's proven reserves of petroleum.

As compared to oil or gas resources, coal reserves are often characterized as widely dispersed. On the one hand, this is an accurate characterization, because major portions of the global reserve base are spread among the continents. On the other hand, the eight nations listed in Table 1.1 hold 88 percent of reported proven recoverable reserves. Leading this list is the United States, with proven recoverable coal reserves of about 270 billion tons.[1]

Table 1.1
Nations Dominating Reported Reserves of Coal

Top Coal-Reserve Holdings	Percentage of Global Reserves	Percentage of 2006 Tonnage Produced
United States	27.1	17.0
Russia	17.3	5.0
China	12.6	38.4
India	10.2	7.2
Australia	8.6	6.0
South Africa	5.4	4.1
Ukraine	3.8	1.3
Kazakhstan	3.4	1.6
Top eight nations	88.4	80.6

SOURCE: BP (2007).

---

[1] When applied to U.S. coal resources, the phrase *proven recoverable reserves* corresponds to what the Energy Information Administration (EIA) terms *recoverable reserves* (EIA, 2006d).

Recent estimates place 2006 global coal production at 6.2 billion tons per year. The energy content of this quantity of coal is about 125 quads,[2] which is nearly 80 percent of the energy content of annual oil production worldwide. As shown in Table 1.1, the eight countries with the largest coal reserves account for more than 80 percent of world production, with China and the United States dominating the list. In 2006, the United States produced a record 1.16 billion tons of coal, which had a total energy content of 23.8 quads.

Most coal is used in the same country in which it is produced. The international coal trade represents about 15 percent of global production and is dominated by demand from Japan, South Korea, and Taiwan. The principal use of coal is to generate electric power. In highly developed economies, nearly all coal is used for power production. For example, more than 92 percent of U.S. coal consumption in 2006 was dedicated to electric power generation (EIA, 2008d, Table 7.3). In less developed economies, power generation is still the dominant application. However, in some cases—e.g., China—a much higher fraction of coal is consumed to support industrial production or to heat commercial and residential buildings.

The Adequacy of the U.S. Coal Resource Base

Is the U.S. coal resource base sufficient to support a domestic CTL industry? To address this question, we hypothesize a large and strategically significant level of CTL production—namely, three million barrels, which is about 15 percent of current petroleum consumption and close to the maximum amount that could be produced by 2030, as shown in Chapter Three. As will be discussed in more detail in that chapter, producing one barrel of coal-derived product requires mining slightly less than 0.5 tons of coal. Producing three million bpd will require mining an additional 550 million tons of coal per year. Over 100 years, for example, this level of mining would consume about 55 billion tons of the 270 billion tons that are reported as the proven coal reserves of the United States.[3]

Whether 55 billion tons of coal are available for CTL production depends on three issues: first, competing demands for coal (namely, for electricity production); second, the accuracy of the estimate of recoverable proven reserves; and third, the extent to which current and future environmental regulations limit coal mining. The first two of these issues are addressed in this section. The environmental constraints on future coal production are discussed in Chapter Six.

---

[2]   A quad is a quadrillion (i.e., 10^{15}) British thermal units (Btu). One quad equals 1.06 exajoules. An exajoule is 10^{18} joules.

[3]   The 55-billion-ton coal requirement is based on the estimated conversion efficiency using current CTL technology. Technical progress is likely to shift this 100-year requirement to between 45 billion and 50 billion tons. Consequently, our estimate of 55 billion tons should be considered as a worst-case upper bound.

Competing Demands for Coal in the United States

Each year, the Energy Information Administration (EIA), part of the U.S. Department of Energy, publishes a series of projections of energy supply and demand in the United States. For coal, EIA's *Annual Energy Outlook 2008* (EIA, 2008c, Table 15) projections show that coal demand through 2030 will be primarily driven by growth in the demand for electric power and by the competitiveness of coal compared to natural gas as a fuel for generating power. For EIA's reference-case projection, annual coal production for traditional uses (i.e., not for liquid-fuel production) is predicted to rise from current levels of 1.16 billion tons to 1.43 billion tons.[4]

Assuming that coal demand for traditional uses stabilizes at the 2030 level projected by EIA, adding three million bpd of CTL capacity would require an annual coal production level of about two billion tons. This rate of production would deplete reported proven coal reserves within 135 years.

However, EIA's projection of coal demand for traditional uses is problematic. According to the rules that guide the development of the estimates published in the *Annual Energy Outlook* series, the projections do not incorporate the possible effects of new legislation that might be enacted to reduce U.S. emissions of greenhouse gases. A consequence of any reasonable legislation directed at significantly reducing greenhouse-gas emissions will be a substantial increase in the price of electricity and a shift away from coal to power-generation approaches that produce lower carbon dioxide emissions.[5] This will be the case whether the legislation calls for a carbon cap-and-trade system or a carbon tax, or it mandates specific low-carbon technologies for power generation.

A number of recent studies have examined coal-demand projections in a carbon-constrained world. A 2006 EIA analysis of alternative greenhouse-gas reduction goals predicts that meeting any appreciable greenhouse-gas reduction goals[6] will cause 2030 coal production to be less than 1.2 billion tons (EIA, 2006b). A more recent EIA analysis of proposed legislation (S.280, the Climate Stewardship and Innovation Act of 2007) found that 2030 coal use for power generation could be significantly below current levels, in some cases less than half (EIA, 2007c). Similarly, modeling work reported in a recent Massachusetts Institute of Technology (MIT) study showed that regulatory measures capable of significantly reducing the carbon dioxide emissions associated with electric power generation will also lead to marked reductions in coal

---

[4] In its reference-case projection, EIA estimates 2030 coal-mine production at 1.46 billion tons. This case also allocates about 30 million tons of coal production to CTL plants in 2030.

[5] We list these two impacts because they are directly relevant to the argument we make in this section. Other outcomes would likely include the development and implementation of carbon sequestration and changes in the way energy is used and priced throughout the economy.

[6] Specifically, goals allowing no more than a 15-percent increase in 2030 greenhouse-gas emissions, as compared to 2004 emissions, resulted in a prediction that 2030 U.S. coal production would be below 2004 levels.

use in the United States compared to what would have been the case in the absence of such control measures (MIT, 2007).

Under the assumption that the U.S. government will eventually adopt effective measures to control carbon dioxide emissions, a more reasonable projection of the upper range of potential 2030 domestic coal production to meet traditional demands (primarily electricity generation) is between 1.1 billion and 1.4 billion tons per year.[7] Adding 0.6 billion tons to support a three million bpd CTL industry increases total domestic production to between 1.7 billion and two billion tons per year. An annual rate of production within this range would deplete the nation's proven coal reserves of 270 billion tons within 135 to 160 years.

Quality of U.S. Coal-Reserve Estimates

In a recent review of the quality of available information on the coal resource base, the National Research Council found that U.S. coal-reserve estimates are based on "old and out-of-date data" and on estimation "methods that have not been reviewed or revised since their inception in 1974" (NRC, 2007, p. 4). The National Research Council study raised the possibility that actual recoverable reserves might be significantly less, especially when taking "into account the full suite of geographical, geological, economic, legal, and environmental characteristics" (NRC, 2007, p. 4). While such a possibility might be realized, it is more likely that an updated assessment of U.S. coal resources will result in a significantly larger estimate of proven coal reserves. This judgment is based on the following considerations:

- There is considerable room for growth because the current estimate for proven coal reserves represents less than 3 percent of the U.S. Geological Survey (USGS) estimate of total coal resources that exist within the United States.[8]
- Advances in coal mining and mine reclamation technology, along with the increases in coal prices that have already occurred since 1974 and that are likely to continue, should enable greater portions of the overall resource base to be mined economically and with reduced environmental impacts.
- There are known cases in which state-level information or actual mining experience shows the existence of appreciable minable resources that are not included as proven coal reserves (SSEB, 2006; NRC, 2007).

---

[7]   This upper range assumes the timely development and commercial application of technology for capturing and sequestering carbon dioxide emissions produced by power plants.

[8]   The USGS estimate of total U.S. coal resources includes four trillion tons of both identified and undiscovered resources from the USGS 1974 estimate and 5.5 trillion tons of mostly undiscovered resources in Alaska (Flores, Stricker, and Kinney, 2004). To further support this room-for-growth point, the principal author of the 1974 estimate of U.S. coal resources, Paul Averitt, emphasized that "the estimates of identified resources . . . are still minimal estimates" (Averitt, 1981, p. 62).

Considering that the passage of time will likely expand the proven coal-reserve base, it is likely that U.S. coal reserves can support more than 100 years of coal demand for electricity generation as well as a domestic CTL industry capable of producing strategically significant amounts of liquid fuels—at least three million bpd.

As discussed in Chapter Three, building a CTL industry capable of approaching such a high level of production would take many decades. This gives adequate time for a major reassessment of the nation's recoverable coal reserves, as called for by the National Research Council (NRC, 2007). Meanwhile, USGS has under way a National Coal Resource Assessment project, which is described as "a multi-year effort . . . to identify, characterize, and assess the coal resources that will supply a major part of [U.S.] energy needs during the next few decades" (USGS, undated). Our discussions with USGS and the U.S. Department of Energy personnel involved with this assessment indicate that it is being conducted at a fairly low level of effort. Correcting the deficiencies noted in the National Research Council report and providing a more reliable estimate of the national coal resource base will require a major increase in resources and number of personnel that is well beyond the scope of the current USGS assessment.

The Distribution of U.S. Coal Reserves and Production

CTL plants will likely be located near coal mines, because it is generally less expensive to transport the liquid-fuel products to customers than to transport coal to CTL plants located near demand centers for liquid fuels. Some CTL plants may be dual-feed plants, consuming both coal and biomass, such as agricultural or forest residues or nonfood crops (e.g., switch grass) especially grown for their energy value (see Chapter Three). In general, biomass resources are more expensive than coal to transport, and site selection would be based on minimizing the overall delivered costs of both feedstocks.

Coal resources are broadly distributed throughout the United States. Coal mines are operating in 26 states. Recoverable reserves are located in 33 states, of which the 16 states listed in Table 2.1 account for about 97 percent of the nation's total (EIA, 2006d, Tables 1 and 15).

While recent coal production has been at an all-time record high, over the past 20 years, there has been a shift in production from states located in the Midwest and the Appalachian region to the western states and particularly to the Powder River Basin in Wyoming and Montana. This geographic shift has been accompanied by greater reliance on large surface-mining operations, mainly due to the geological characteristics of western coal deposits and technical advances that have lowered the costs of surface excavation of coal. Currently, about 58 percent of all U.S. coal production comes from fewer than 100 mines in the West. The remaining 42 percent derives from a mix of

Table 2.1
Recoverable Coal Reserves and 2005 Coal Production by State

State	Recoverable Coal Reserves (billions of tons)	2005 Coal Production (millions of tons)
Montana	74.9	40
Wyoming	40.6	404
Illinois	38.0	32
West Virginia	18.0	154
Kentucky	14.9	120
Pennsylvania	11.8	67
Ohio	11.5	25
Colorado	9.8	39
Texas	9.5	46
North Dakota	6.9	30
New Mexico	7.0	29
Indiana	4.1	34
Virginia	3.1	28
Alaska	2.8	1
Alabama	2.8	21
Utah	2.7	25
16-state total	258.4	1,095
U.S. total	267	1,131

SOURCE: EIA (2006d, Tables 1 and 15).

nearly 1,300 surface and underground mines located east of the Mississippi. The 16 states listed in Table 2.1 account for 97 percent of U.S. coal production.

Coal Variability

Coal is a complex substance. Its composition and characteristics vary greatly among the deposits found within the United States and in other countries. This is due to variation in its botanic origins. While coal is often described as a mineral, a more accurate descriptor is *organic rock*. But compared to most other rocks, coal is very light. Its specific gravity is only slightly higher than that of water. With regard to CTL production, the key variable is the coal's *rank*.[9] However, even within the same rank, the ash content and agglomeration properties during heating can be decisive in process design (Elliott and Yohe, 1981).

The four ranks of coal found in the United States are, from lowest to highest, lignite, subbituminous, bituminous, and anthracite. The higher-ranking coals are generally older, harder, and brighter, and they also have higher heating values. Even within a rank, there can be considerable variation, as shown in Figure 2.1.

The most abundant and widespread coal in the United States is bituminous coal. Nearly all the recoverable coal resources, as well as coal currently being produced east of the Mississippi River, are bituminous. Additionally, notable bituminous coal reserves and production occur in Colorado, New Mexico, and Utah. Subbituminous coals are also abundant in the United States. Their recoverable reserves and production are limited to six states: Wyoming, Montana, New Mexico, Colorado, Washington, and Alaska. Lignite is found primarily in North Dakota, Texas, and Montana, but there are also active lignite mines in Louisiana and Mississippi. Anthracite resources are concentrated in northeastern Pennsylvania. Extremely important in the past, U.S.

Figure 2.1
Approximate Heat Content of Different Ranks of Coal

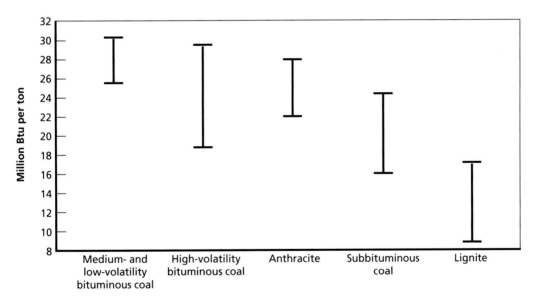

SOURCE: EIA (1995).
NOTE: This figure depicts the heat content of as-received coal, which includes natural moisture and combustible and incombustible materials.
RAND MG754-2.1

---

9   The key factors in establishing rank are carbon content and calorific value (e.g., Btu per pound).

anthracite resources are now nearly depleted. Recoverable anthracite reserves are estimated at 760 million tons. Current annual production is less than two million tons, a sizable fraction of which is from waste piles accumulated prior to the 1950s.

As a commercial CTL industry develops in the United States, operating experience on various types of coal will build the empirical knowledge base required to lower the technical risks of designing and operating CTL plants. Early CTL plants will likely compensate for the technical uncertainties associated with coal variability by incorporating conservative designs and operating procedures for components that receive and initially process coals.

Mine Size

As discussed in Chapter Three, full-scale CTL plants will produce between 30,000 and 80,000 bpd of liquid fuels. Such large plants will each consume 15,000 to 40,000 tons of coal per operating day or, equivalently, between 6.5 million and 13 million tons per year. Over a 50-year operating lifetime, each commercial CTL plant will consume between 300 million and 700 million tons. To be competitive, these plants will need to be sited so that the costs of delivering coal to them will be low. For plants built in the West, subbituminous and lignite deposits are sufficiently concentrated and extensive to allow each plant to be served by a single large mine. For example, 13 mines currently operating in the West are each producing more than 12 million tons per year (EIA, 2006d, Table 9).

East of the Mississippi, only five mines are annually producing more than 6.5 million tons. In this region, a few CTL plants might be sited so that they are supplied by a single mine, but it is more likely that a single plant will require the development of two or more coal mines along with dedicated coal-delivery systems. It is reasonable to anticipate that, if economies of scale are balanced with coal-delivery costs, the CTL plants built in the Midwest and in Appalachian regions will be smaller than those built west of the Mississippi.

Policy Implications of the Coal Resource Base

The world's proven recoverable reserves of coal are very large, nearly triple the energy value of the world's proven reserves of petroleum. The United States leads the world in the amount of reported recoverable reserves of coal. The other leading countries, with respect to the reported size of the reserve base and current production, are Russia, China, India, Australia, and South Africa. Using a portion of these coal reserves to produce liquid transportation fuels would diversify the liquid-fuel supply base and reduce demand from traditional suppliers of crude oil. Chapter Five further develops the economic and national security consequences of reducing global oil demand.

Considering the geographic and geologic disposition of U.S. coal resources, recoverable coal reserves are sufficient to support a CTL industry producing a few million bpd and to continue to fuel electric power generation. Dedicating 15 percent of recoverable U.S. coal reserves to CTL production would yield roughly 100 billion barrels of liquid transportation fuels. As a CTL industry develops, it would be prudent to reassess the nation's coal-reserve base and the effect on coal demand from measures that may be forthcoming for the purpose of reducing greenhouse-gas emissions in the United States. Concerns with the environmental and safety impacts of coal mining may impede or even prevent the development a large CTL industry in the United States, as discussed in Chapter Six.

Within the United States, substantial coal resources are located in at least 16 states. The disposition of coal reserves in the West—such as those in Montana, Wyoming, and North Dakota—allows relatively low mining costs. Coal from these states could be used in large mine-mouth CTL plants or transported to out-of-state CTL plants that would use coal or a combination of coal and biomass. The latter approach might allow the CTL plant to access more abundant and lower-cost water supplies. In the eastern and central regions of the United States, very large CTL plants are less likely, since the logistics of moving coal from multiple mines to a central facility may impose unacceptable costs. The central and eastern regions of the United States also have greater levels of rainfall, which promotes higher yields of biomass, thereby enhancing the prospects of CTL plants that accept both coal and biomass.

The variability of coal has consequences for the development of a commercial CTL industry in the United States. The fact that nonagglomerating bituminous coals from South Africa have successfully supported commercial CTL operations in that nation does not justify the claim that CTL is a *commercial* technology for applications outside of South Africa. Rather, a residual of technical uncertainty remains regarding the design and performance of plants intended to operate on various types of U.S. coals. For initial plants, this problem is handled by conservative (and more expensive) designs and conservative (i.e., reduced) performance expectations, both of which are barriers to investment. Only through the construction and operation of a few CTL plants will the design community build the knowledge base required to correlate with confidence coal properties with process design and performance.

CHAPTER THREE

Coal-to-Liquids Technologies

This chapter provides an overview of the principal technical approaches for producing liquid fuels from coal and describes the fuels produced. It also examines the current status, and prospects for, commercial CTL operations and discusses methods of controlling greenhouse-gas emissions. Finally, the chapter reviews production costs and establishes a minimum timeline for the buildup of liquid-fuel production from a commercial industry.

Current interest in coal-derived liquid fuels has, for the most part, concentrated on approaches that begin with coal gasification and fall into a technical category known as *indirect liquefaction*. The best-known of these indirect methods is Fischer-Tropsch (FT) liquefaction. The technology for the FT approach is the key focus of this chapter. This chapter also reviews the other major indirect liquefaction method that is ready for near-term applications—the methanol-to-gasoline (MTG) CTL process.

The other principal CTL approach is hydroliquefaction, which is often referred to as *direct liquefaction*. During the 1970s and 1980s, the U.S. government made very large investments in developing direct liquefaction, but the technology has not been commercially applied since World War II. However, a large commercial direct-liquefaction plant is now under construction in China. While the direct liquefaction process is less familiar, its status is also reviewed in this chapter because it may be a viable alternative.

The Fischer-Tropsch Coal-to-Liquids Approach

Overview of the Process

A highly simplified schematic of the FT CTL process is shown in Figure 3.1. The process begins with the gasification of coal, which consists of reacting coal with steam and oxygen at elevated temperatures (1,000 to 1,500 degrees Celsius) and moderate pressures (~500 pounds per square inch [psi]) to produce a mixture of hydrogen, carbon

Figure 3.1
Simplified Process Schematic for Fischer-Tropsch Coal-to-Liquids Systems

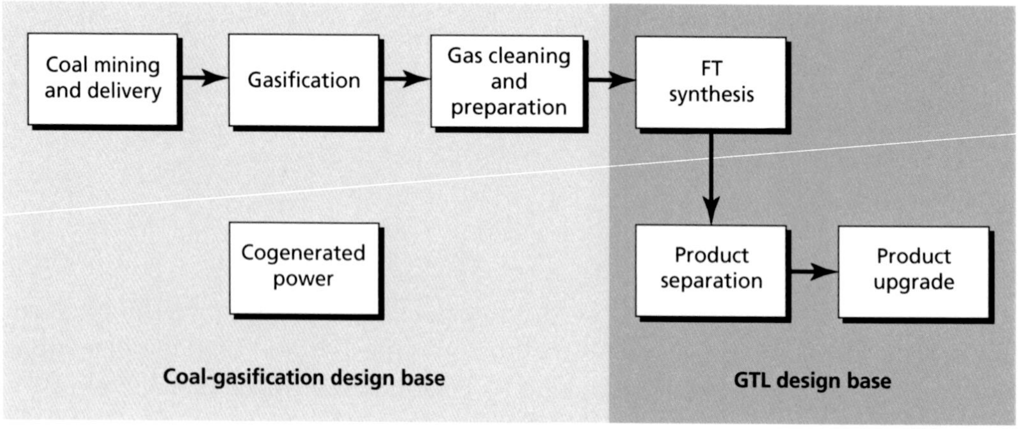

NOTE: GTL = gas to liquids.
RAND *MG754-3.1*

monoxide, and carbon dioxide.[1] A gas consisting mainly of the first two of these constituents is called *synthesis gas*. But as it leaves the gasifier, the gas is dirty: It contains
carbon dioxide and various gaseous molecules that derive from the impurities found
in coal. These impurities would harm the performance of subsequent processing steps
and are therefore removed in sections of the FT CTL plant that are designed to clean
and properly prepare the synthesis gas.[2] Sulfur compounds are reduced to near-zero
concentrations. In general, the captured sulfur would be converted to pure solid sulfur
or sulfuric acid, both of which are articles of commerce as opposed to wastes. It is also
during gas cleaning that extensive removal of trace mercury compounds would occur.

A consequence of gas cleaning and preparation is a highly concentrated stream
of carbon dioxide. In the absence of a greenhouse-gas management requirement, this
carbon dioxide would be released into the atmosphere.

The next step is to send the cleaned synthesis gas to FT reactors, where it is cat-
alytically converted to a mixture of hydrocarbons.[3] This mixture generally includes

---

[1]   In a gasifier, the desired reaction is between coal and steam, which yields the fundamental components of
synthesis gas: carbon monoxide and hydrogen. However, the reaction is highly endothermic (i.e., it absorbs heat).
In commercial gasifiers that are candidates for CTL applications, the requirement for heat is met by introducing
oxygen into the gasifier so that some of the coal is combusted. This is the primary source of the carbon dioxide
leaving the gasifier.

[2]   Depending on the catalyst used in the FT synthesis reactor, preparation of the synthesis gas may require raising the gas's hydrogen content. This involves reacting a portion of the carbon monoxide in the synthesis gas with
steam, which results in additional production of carbon dioxide.

[3]   Significant amounts of carbon dioxide are also produced in certain types of FT reactors—namely, those
with catalysts, such as iron-based catalysts, that do not require that the hydrogen content of the synthesis gas be
shifted.

hydrocarbon gases, such as methane and propane; hydrocarbons that are typically found in gasoline, diesel, and jet fuel; and heavier compounds that are categorized as waxes.[4] These various streams are separated, primarily according to their boiling points, and can be further treated to produce two product streams: naphtha and middle distillates.[5] At the FT CTL plant, the middle-distillate product can be retail-ready diesel fuel or a combination of diesel fuel and jet fuel. The naphtha product is basically a very low–octane (i.e., about 40 octane) gasoline that must be extensively upgraded before it can be used as an automotive fuel. Alternatively, the naphtha can be converted to ethylene and related chemical feedstocks. At a large CTL plant, this upgrading to gasoline could take place at the plant site, but, for smaller CTL plants, the naphtha would likely be sent to a refinery or to a central processing facility built to serve two or more CTL plants.

Extensive amounts of heat are generated at various stages of the FT CTL plant. In particular, a large amount of heat is liberated in the FT reactors, since the conversion of synthesis gas to hydrocarbons is highly exothermic (i.e., it releases heat). Substantial heat may also be released when hot, dirty synthesis gas leaving the gasifier is cooled to the near-ambient temperatures needed for the gas-cleaning steps. Much of this heat is captured and used to cogenerate electric power at the FT CTL plant. In general, more than half of the cogenerated power will be used within the FT CTL plant, primarily to power the air-separation units required to produce oxygen for the gasifiers.

The energy efficiency of an FT CTL plant depends on the details of the plant design. Important factors determining energy efficiency are what type of coal is used,[6] which components are selected (especially the gasifiers and cooling towers), and how the plant design provides for recovery and management of heat. Recent engineering analyses of FT CTL plants show that those designed to operate on bituminous and subbituminous coals should have an overall energy efficiency of close to 50 percent (NETL, 2007b; SSEB, 2006, Appendix D).[7] This estimate is for first-of-a-kind moderate-sized or larger plants, i.e., plants producing at least 25,000 barrels of liquid products per day. Smaller FT CTL plants may fall slightly below this range because heat losses tend to increase with decreasing equipment sizes.

These values for energy efficiency are based on the total product slate of the FT CTL plant, including electricity sold to the grid. Considering only liquid produc-

---

[4]   Waxes are composed of predominantly straight-chain hydrocarbons having more than 20 carbon atoms.

[5]   Some of the hydrocarbons leaving the FT reactor have a market value above that of fuels, and the first few FT CTL plants might attempt to extract them for sale as feedstocks for chemical plants. However, if a large FT CTL industry is established, the market for these by-products will become saturated.

[6]   In general, lignites have high moisture levels, and the need to evaporate this additional moisture causes energy efficiency to decrease.

[7]   All energy efficiencies reported herein are based on the higher heating value of the fuels, which is the general practice when dealing with coal and liquid fuels.

tion, the energy efficiency is about 52 percent.[8] Capturing carbon dioxide, as discussed later in this chapter, reduces overall plant efficiency by one to two percentage points, depending on the extent of removal.

FT CTL plant energy efficiencies in the 50-percent range underlie the rule-of-thumb estimate that one ton of bituminous coal yields two barrels of FT liquids.[9]

Development Status

During the 19th century, the gasification of coal became widespread in both Europe and the United States. The purpose was to produce *town gas*, which was used primarily for lighting and cooking. In the 1920s, two German scientists, Franz Fischer and Hans Tropsch, discovered a method of converting the principal constituents of town gas—hydrogen and carbon monoxide—into liquid fuels.

Beginning in late 1933, Germany's Nazi government promoted the development of a CTL industry to secure a domestic supply of automotive fuel. German production of FT CTL peaked in 1944 with nine plants producing 4.2 million barrels, a per-plant average of about 1,300 bpd.

In the 1950s, concerned that a trade boycott could cut off petroleum imports because of its apartheid policy, the government of South Africa decided to subsidize the production of transportation fuels from its domestic coal resources. Today, Sasol in South Africa operates the world's only commercial CTL plant. This single plant derives from the integration and upgrading of two plants built by Sasol in the early 1980s, and it currently produces the energy equivalent of about 160,000 bpd of fuels and chemicals (IEA Coal Industry Advisory Board, remarks by Andre Steynberg, 2006; Sasol, 2006).

Shared Technology Base. While commercial-scale experience with FT CTL is extremely limited, the technology base has moved considerably forward, especially over the past 15 years. This is because the front end (i.e., gasification and gas cleaning) of an FT CTL plant uses much the same technology as do other systems that involve coal gasification, and the back end uses much the same technology as do modern GTL plants, as shown in Figure 3.1.

According to a worldwide survey, at least 27 new facilities based on coal gasification began operations between 2000 and the end of 2007 (NETL, 2007f). Nearly all of these coal-gasification facilities are devoted to producing synthesis gas for the manufacture of chemicals, mainly ammonia and methanol. Three facilities are dedicated to producing electric power using a combination of gas and steam turbines that is often

---

[8] Liquid-only energy efficiency is calculated by assuming that the electric power sold by a CTL plant would have otherwise been produced by a highly efficient (i.e., heat rate of 8,000 Btu per kilowatt-hour [kWh]) coal-fired power plant. Energy input for liquid-only production is determined by subtracting this amount of coal from the total coal feed of the plant.

[9] Subbituminous coals have a lower energy content and consequently yield slightly less oil, about 1.8 barrels per ton, than do bituminous coals.

referred to as an integrated gasification combined cycle (IGCC). Coal-gasification facilities, whether for chemicals or power, involve much the same operations as would be required at the front end of a modern FT CTL plant—namely, preparing and feeding coal to a pressurized gasifier, deeply cleaning the synthesis gas, removing and handling the ash or slag rejected by the gasifier, managing heat transfers, and, in some cases, shifting the carbon-monoxide-to-hydrogen ratio and removing carbon dioxide.

A favorable attribute of the FT approach to liquid-fuel production is that synthesis gas can be produced from a variety of feeds, including natural gas, petroleum coke, biomass, and, of course, coal. Over the past 15 years, commercial interest in FT technology has centered on stranded deposits of natural gas.[10] In the early 1990s, Shell built a pioneer GTL plant at Bintulu, Malaysia. This plant continues to operate and is currently producing about 14,700 bpd of liquids. In 2004, Sasol converted its oldest FT CTL plant to an FT GTL plant using pipeline natural gas from Mozambique. Two GTL plants, one using Sasol technology and the other using Shell technology, with a combined output of at least 240,000 bpd, are being constructed at Ras Laffan Industrial City in Qatar.[11] The initial phase of the Qatar-Sasol plant began startup operations in late 2006 with a rated capacity of 34,000 bpd. Several other GTL projects are in development or in the early stages of construction.

A consequence of the increasingly favorable commercial prospects of FT GTL technology has been significant private-sector investment toward development of improved technology for FT conversion and the downstream upgrading of FT products. Leaders in this area include Shell, Sasol, Chevron, and Exxon Mobil, as well as a few smaller firms, most notably Rentech and Syntroleum Corporation, both of which are U.S. firms.

Planned Fischer-Tropsch CTL Projects. A fairly large number of projects to build FT CTL plants have been announced. However, as of December 2007, none had begun construction. In the United States, front-end engineering designs[12] are reportedly under way for three projects: a project by Rentech to produce 1,600 bpd of FT

---

[10] *Stranded* refers to natural-gas deposits that pipelines cannot easily bring to large markets. Marketing these resources requires either cryogenically converting and transporting the natural gas as liquefied natural gas (LNG) or converting it via a GTL process to liquid fuels that are easily transportable.

[11] Both projects are joint ventures with Qatar Petroleum. The Sasol plant, known as Oryx GTL, is in operation. Sasol Chevron is pursuing a 65,000-bpd-capacity addition. The Shell plant, the Pearl GTL project, is designed to produce 140,000 bpd of diesel, naphtha, and liquefied petroleum gas (LPG).

[12] As used in this report, *front-end engineering design* refers to a site-specific engineering design conducted at a level of detail that will provide a confident prediction of capital and operating costs, construction requirements, and environmental performance. The front-end engineering design serves as the basis for selecting an engineering, procurement, and construction (EPC) contractor that will be responsible for most of the subsequent engineering (i.e., detailed design). The environmental analyses that are part of the front-end engineering design activity provide the basis for selecting environmental control systems and obtaining preconstruction permits.

liquids that will be located near Natchez, Mississippi;[13] a project by Baard Energy to produce 50,000 bpd of FT liquids in Wellsville, Ohio;[14] and a project by WMPI to produce 5,000 bpd of FT liquids in Frackville, Pennsylvania.[15] To our knowledge, no other FT CTL projects in the United States or elsewhere have advanced to the front-end engineering design stage of design.

Fischer-Tropsch Products

Scientists studying FT synthesis have determined that a polymerization reaction produces chains of carbon atoms. The length of these chains follows a distribution curve that governs the fractions of light gases, naphtha, diesel, and waxes that will be produced. Using the FT method to produce liquid fuels, it is not possible to avoid generating a fairly broad mix of products (Wender and Klier, 1989).[16] By changing catalysts and operating conditions, the distribution can be shifted but never avoided. For example, an FT reactor can be run to optimize gasoline and reduce diesel production, but this operating regime results in very high yields of hydrocarbon gases, especially methane. To avoid methane, recent developments in FT fuel production have focused on increasing the fraction of wax, which can be cracked to produce useful transportation fuels.

Because of the chemical mechanism underlying FT synthesis, most of the hydrocarbons leaving the FT reactor are classified as paraffins, and nearly none are aromatics.[17]

Fischer-Tropsch Diesel Fuel. The middle-distillate fraction produces an exceptionally high-quality diesel fuel. Sulfur levels are less than one part per million (ppm), well below the current U.S. Environmental Protection Agency specification of 15 ppm.

---

[13] Rentech plans to build the Natchez CTL facility in two phases. The first phase, for which front-end engineering design is under way, would demonstrate Rentech's first commercial-scale FT reactor. The facility would be designed to gasify either coal or petroleum coke in combination with a small amount of biomass. The first phase also includes capturing and selling carbon dioxide, which would be used for enhanced oil recovery. A feasibility study is under way for a potential second phase that would produce an additional 28,000 bpd. There is a possibility that a third phase would also be built.

[14] The Baard Energy project seeks to produce ultraclean transportation fuels through the gasification and processing of domestic coal in conjunction with biomass cofeed, carbon-capture-and-storage technology, and combined-cycle cogeneration processes to reduce carbon dioxide emissions.

[15] The WMPI project would convert waste anthracite, which is widely distributed in eastern and central Pennsylvania, to FT liquids using technology licensed from Shell, Sasol, and Chevron. This project was conceived in the mid-1990s and has been partially supported by funds awarded by the U.S. Department of Energy. An environmental-impact statement has also been completed for this project.

[16] The exception to this rule is that single-carbon chemicals (namely, methane and methanol) can be synthesized with near-100-percent selectivity.

[17] Paraffins are fully saturated hydrocarbons. Generally, they burn with a clear yellow or bluish flame. A familiar example is candle wax. Aromatic hydrocarbons consist of one or more benzene-like rings. Aromatics tend to burn with a smoky, sooty flame (Mohrig, Hammond, and Schatz, 2006) and are frequently classified as carcinogens.

Because FT diesel has a high paraffin content and a near-zero aromatic content, its cetane number is very high. The cetane number is a measure of how readily diesel fuel ignites. The higher the cetane number, the sooner a fuel will start to burn after it is injected into a combustion chamber. Coal-derived fuels from the FT process will generally have a cetane number from 70 to 80. This is significantly higher than that of refinery diesel fuel, which generally ranges from 40 to 55.

In general, fuels with higher cetane numbers make starting a cold engine easier and reduce hydrocarbon and soot pollutants generated in the minute or so following a cold start. Higher cetane-number fuels also tend to reduce nitrogen oxide and particulate emissions from a warm engine, although the amount of such reductions is dependent on engine design (Norton et al., 1998; Clark et al., 1999; May, 2003; Maly, 2004; Schaberg, 2006).

These beneficial performance characteristics could provide FT diesel with a premium price as a blend stock with lower-grade conventional diesel. This would allow conventional-petroleum refiners to reduce the amount of hydrotreating required to meet federal specifications for sulfur content while maintaining required cetane levels.

Fischer-Tropsch Jet Fuel. Aircraft refueling at Johannesburg's O. R. Tambo International Airport routinely receive a blend of FT-derived and conventional jet fuel. This practice is based on standards that were established for blends of up to 50 percent of a Sasol-produced FT-derived jet fuel and approved in 1999 by a consensus of the international aviation industry (Chevron Global Aviation, 2006).[18] This standard-and-approval process is now being extended to address generic blends (i.e., blends based on any FT-derived jet fuel) and to aviation use of a fully synthetic jet fuel formulated from FT products (FAA, 2008; Altman, 2007).

Fuel and engine testing, experience with the Sasol jet blend, and progress reports covering the ongoing standard-and-approval process indicate that FT-derived jet fuel will be usable in blends with Jet A or Jet A-1 fuels.[19] The issue is not whether but rather how such fuels should be formulated—namely, under what specifications for ensuring safety and reliable performance.

In 2006, the U.S. Air Force began a program to certify that a 50/50 blend of FT and the principal military jet fuel, JP-8, is safe for operational use in military aircraft. Flight tests of an FT/JP-8 blend in a B-52H, C-17, B-1, F-15, F-22, and KC-135 have been completed, and the blend has been certified as safe for use in B-52H, C-17, and B-1 aircraft.[20]

---

[18] The approval process culminates in listing by ASTM International (formerly the American Society for Testing and Materials) and the UK Ministry of Defence.

[19] Jet A-1 is the jet-fuel formulation used in most of the world by commercial aviation. It is very similar to Jet A, which is used primarily in the United States.

[20] Chapter Eight provides additional information on this U.S. Air Force program.

Two concerns have been raised regarding the use of a commercial or military jet fuel derived solely from FT products. Certain elastomers used as seals in aircraft fuel systems swell on exposure to aromatic hydrocarbons. Replacing conventional jet fuel with a low-aromatic FT jet fuel will cause these elastomers to shrink, possibly resulting in a fuel leak. The susceptible elastomers are in the nitrile polymer family and are found only in older-model aircraft. The second concern is that the volumetric energy density of a pure FT jet fuel is about 3.7 percent lower than that of petroleum-derived jet fuels.[21] Because the maximum fuel loading of an aircraft is determined by the volume of its fuel tanks, aircraft powered by FT jet fuel will have a reduced maximum flight range. However, only a small fraction of aircraft operations would suffer from this limitation.

Research on and testing of FT fuels in jet engines reveal improved thermal stability and significantly reduced particulate emissions as compared with conventional or military jet fuel (Edwards et al., 2004). Higher thermal stability results in reduced carbonaceous deposits in fuel systems and should lower maintenance requirements and improve performance. However, the economic value of these improvements has not yet been established.

The International Civil Aviation Organization has established limits for emissions of nitrogen oxides, carbon monoxide, unburned hydrocarbons, and smoke from commercial jet engines during takeoff and landing (Chevron Global Aviation, 2004). At present, these limits are being met through improved engine designs and have not resulted in any change in the specifications governing jet fuels. Sulfur emissions from aviation are not controlled. The current sulfur specification is 3,000 ppm, although most jet fuel is between 500 and 1,000 ppm (Chevron Global Aviation, 2004). While the use of a FT-derived jet fuel or blend is likely to cause significant reductions in both particulate matter and sulfur emissions, FT products are unlikely to command a premium price over conventional jet fuel without more stringent limits on aircraft emissions.

Fischer-Tropsch Naphtha. Between 20 and 40 percent of the total product from a modern FT-based-fuel plant will be light liquid hydrocarbons classified as naphtha. Somewhat similar product streams are associated with conventional refining of petroleum, although upgrading FT naphtha is more difficult. For the initial round of CTL plants that might be built in the United States, it is likely that the raw naphtha leaving the FT reactor will be hydrotreated and transported to a refinery. A potentially more attractive option, especially for early FT CTL plants, would be to dedicate the raw naphtha stream to chemicals production, which would likely involve some treatment at the FT plant site and transport of naphtha components to appropriate chemicals-production facilities. This chemicals approach may provide a higher value to the naphtha

---

[21] This decrease is outside the allowed range for volumetric density as specified in ASTM D1655 Jet A/Jet A-1 (ASTM, 2008). The 50/50 blend falls within the allowed range.

stream than does gasoline production (Rosborough, 2007), but comparative analyses are not available. For this reason, in the cost analyses of CTL plants conducted as part of this study, we assume that, on a volume basis, naphtha has a value that is 30-percent lower than FT diesel fuel that is fully ready for distribution.

The Methanol-to-Gasoline Coal-to-Liquids Approach

Overview of the Process

The MTG CTL approach consists of three major steps. In the first step, coal is gasified to produce synthesis gas; in the second step, synthesis gas is converted to methanol; and, in the third step, methanol is converted into gasoline. The front end of an MTG CTL plant would be similar to that of an FT CTL plant. As shown in Figure 3.2 and in common with the FT method, the process begins with coal gasification followed by the steps necessary to prepare a clean synthesis gas with a hydrogen-to-carbon-monoxide ratio suitable for methanol synthesis. The second major step, the synthesis of methanol, takes place at moderate temperatures and pressures (200 to 300 degrees Celsius, about 750 psi) and is slightly exothermic. The third step, gasoline production from methanol, releases a large amount of heat. To better control this heat release, methanol is first dehydrated to produce dimethyl ether. Using specially designed catalysts, the dimethyl ether is converted to a mix of hydrocarbons that are very similar to those found in raw gasoline. After upgrading and separations, the liquid-fuel yield of an MTG CTL plant would be about 90-percent gasoline, with most of the remainder being LPG. Both products can be directly distributed to end users. Additionally, a small amount of electricity or fuel gas might be produced and sold.

Figure 3.2
Simplified Process Schematic for Methanol-to-Gasoline Coal-to-Liquids Systems

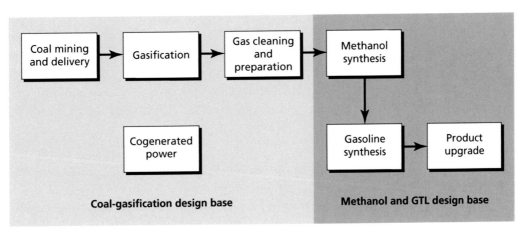

In the early 1980s, the U.S. Department of Energy, through Oak Ridge National Laboratory, sponsored a series of engineering analyses and conceptual designs of MTG CTL plants (Wham et al., 1981; Irvine et al., 1984). More recent design work on these plants is not publicly available. Based on these early designs and taking into account technical advances since then, overall energy efficiencies should be comparable to those of FT CTL plants when operating on the same coal, using the same coal gasifiers, and following similar design strategies for the recovery and management of heat. Specifically, the energy efficiency for liquids production should be in the neighborhood of 50 percent. Capturing carbon dioxide and compressing it for transport would likely lower this figure by a couple of points.

Development Status

MTG CTL is ready for initial commercial applications, because an MTG CTL plant can be constructed using subsystems and components that have been successfully operated at large scales. As shown in Figure 3.2, the front end of an MTG CTL plant is based on coal gasification and synthesis gas cleaning and preparation. As is the case with FT CTL plants, MTG CTL plants can take advantage of the significant advances in coal gasification and gas processing that have occurred over the past 15 years.

Methanol, also known as *methyl alcohol,* is one of the world's major commodity chemicals and is used to produce a large number of consumer and industrial products, including synthetic textiles, plastics, adhesives, and pharmaceuticals.[22] For decades, commercial methanol production, both globally and in the United States, has been based primarily on the catalytic reaction of synthesis gas produced from natural gas. Most importantly, there is ongoing commercial production of methanol using coal-derived synthesis gas in the United States. Specifically, at its Kingsport, Tennessee, facilities, the Eastman Chemical Company is gasifying a high-sulfur bituminous coal to produce commercial quantities (roughly 100,000 gallons per day) of methanol.[23]

The breakthrough discovery that allowed gasoline manufacture from methanol occurred in the early 1970s at the Mobil Research and Development Corporation (Meisel, 1981), which is now ExxonMobil Research and Engineering Company. Mobil researchers found that the structure and size of pores in certain zeolite catalysts promoted hydrocarbon conversion but limited chain growth, thereby allowing a process that produced hydrocarbons mainly in the gasoline boiling range. Moreover, the catalysts were found to produce a mixture of gasoline constituents, such as olefins,

---

[22] The most familiar methanol product for most consumers is windshield-wash antifreeze, which is primarily a solution of methanol and water.

[23] The Kingsport operation became commercial after a successful demonstration that was cost-shared under the U.S. Department of Energy's Clean Coal Technology Demonstration Program. Eastman Chemical recently announced its intent to participate in further commercial ventures for chemical production via coal gasification (Eastman, 2007).

branched paraffins, and aromatics, that promote higher octane ratings in the final gasoline product.

In the early 1980s, a 100-bpd demonstration plant built in Germany with support from both the U.S. Department of Energy and the German government successfully demonstrated the MTG method.[24] The first and only commercial-scale MTG plant was based on natural gas and built in Motunui, New Zealand, with strong support from the government of New Zealand.[25] With a capacity of 14,500 bpd, the Motunui MTG GTL plant went online in 1985 and operated successfully for approximately ten years (Heinritz-Adrian, Brandl, Zhao, et al., 2007; Heinritz-Adrian, Brandl, Hooper, et al., 2007; Tabak, 2006). The Motunui plant was converted to the production of technical-grade methanol, most likely because of the low prices of crude oil prevailing throughout the 1990s.

Two large-scale MTG CTL projects are under way, both based on licenses from ExxonMobil Research and Engineering. In China, Shanxi Jincheng Anthracite Coal Mining has announced that it will construct a 2,600 bpd plant in Jincheng, Shanxi province. This plant will be part of a larger "pilot-scale" facility that will also include advanced coal gasifiers designed to process anthracite and a methanol plant.[26] The engineering phase of this plant started in October 2006, with an award to Uhde GmbH. The plant is currently under construction.

In December 2007, DKRW Advanced Fuels announced that it has selected the MTG approach for a CTL plant that would produce 15,000 to 20,000 barrels of gasoline per day (DKRW Advanced Fuels, 2007). The plant would be near the town of Medicine Bow, Wyoming. This project is currently in the front-end engineering design stage of design. Plant startup is currently scheduled for 2013.

According to ExxonMobil representatives, both the Jincheng and Medicine Bow plants would include significant process improvements based on operating experience obtained in New Zealand.

Methanol-to-Gasoline Products

Relative to an FT plant, a modest amount of processing is required to produce high-quality, salable products from an MTG plant. This is primarily because the MTG reactor does not produce any of the heavy hydrocarbons generally found in diesel oil, jet fuel, and heavy oils and waxes. Additionally, the raw gasoline from the MTG reactor

---

[24] The 100-bpd demonstration plant received locally produced methanol and successfully demonstrated conversion to gasoline. This constituted a demonstration of the overall process concept, since all other portions (i.e., synthesis-gas production from coal or natural gas and methanol production from synthesis gas) of a complete MTG process were commercial.

[25] The Motunui MTG GTL plant was 75-percent owned by the New Zealand government and 25-percent owned by Mobil.

[26] Jincheng Mining has announced a goal of eventually building a 26,000 bpd CTL plant based on the MTG approach.

has a hydrocarbon composition (branched chains and isomers) that is very close to that of a finished product.

The finished MTG gasoline would be zero-sulfur gasoline with properties that are fully compatible with the existing infrastructure for the distribution of gasoline. In particular, regular and higher-octane gasoline formulation should be possible without adding oxygenates. As compared to conventional gasoline from petroleum, according to available information (Heinritz-Adrian, Brandl, Hooper, et al., 2007), MTG-derived gasoline exhibits equal or enhanced properties (Reid vapor pressure and benzene content) relevant to the protection of air quality.

In addition to gasoline, an MTG reactor also produces hydrocarbon gases that, when pressurized, are liquids at ambient temperatures and that can be sold as LPG. After processing required to make a finished gasoline, between 10 and 12 percent of the energy output of an MTG plant could be LPG. The alternatives to selling LPG are to reform it into synthesis gas and make additional gasoline or to burn it to generate additional electric power.

Whether and how much LPG will be produced will depend on the market and prices for LPG in the vicinity of an MTG CTL plant. Consisting primarily of propane and butane, the LPG produced at an MTG plant can substitute for conventional LPG applications, such as heating and cooking in areas in which natural gas is not available. As is the case with conventional LPG, MTG LPG can also be used as a transportation fuel in vehicles that have been appropriately modified.

The Direct Coal Liquefaction Approach

Overview of the Process

In direct liquefaction, a slurry comprising finely ground coal and a solvent is heated to temperatures above 400 degrees Celsius in a pressurized (around 1,500 psi) vessel containing hydrogen and an appropriate catalyst. Under these conditions, the solid organic material in the coal dissolves, continues to react with hydrogen, and breaks down into smaller molecules (Wu and Storch, 1968).

In the simplified scheme shown in Figure 3.3, the coal dissolves in the coal-liquefaction reactor. The lighter molecules are moved to a separate reactor, in which they are further hydrotreated. The heavier paste (consisting primarily of coal ash, catalyst, and heavy oils) is sent to processing units for further conversion, separation, and removal of ash.[27] Most of the heavy oils and some light gases are generally used to make hydrogen. A portion of the heavy oils is separated out and used to form a slurry with the incoming coal. In many process schemes, a portion of the coal feed entering the direct-liquefaction plant would be sent directly to a gasifier to produce

---

[27] Separation is done at lower pressures so that heavy oils can be separated from distillable liquids.

Figure 3.3
Simplified Process Schematic for Direct Liquefaction (Two-Stage)

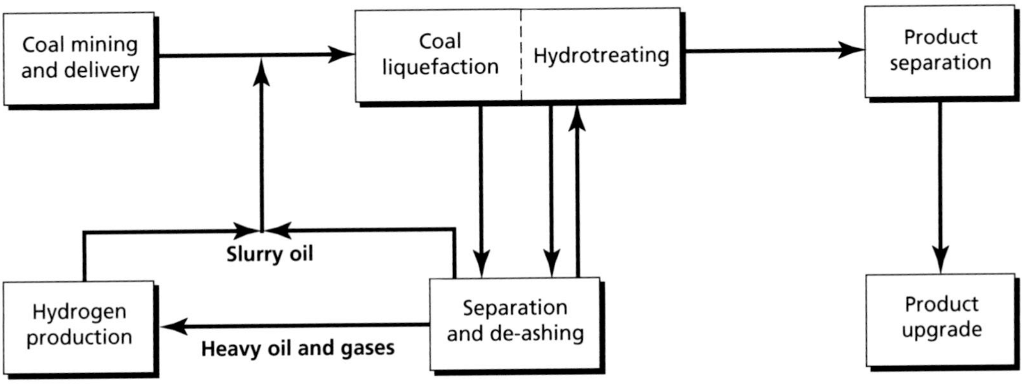

RAND *MG754-3.3*

sufficient hydrogen for the coal-liquefaction and associated hydrotreating operations, as well as the hydrogenation operations that are required for product upgrading.[28]

The principal products of a direct-liquefaction CTL plant would be naphtha and middle distillates. While the properties of these products are highly specific to the direct-liquefaction process by which they are produced, we can make a few general statements about them. Compared to the corresponding streams from normal petroleum, direct-liquefaction products contain more aromatic and cyclic hydrocarbons and fewer paraffins, and they have a lower overall hydrogen content. For these reasons, significant upgrading is required before the products can be used. This upgrading can occur at the direct-liquefaction plant, or the raw products can be sent to a refinery that is specially equipped for the necessary upgrading steps.

Liquid yield from direct liquefaction is very process dependent, but, in general, yields are projected to be considerably higher than those of FT CTL plants designed with current commercial technology. Based on conceptual designs of two-stage and catalytic multistage direct-liquefaction CTL plants, prospective yields are reported to be between 2.7 and 3.0 barrels per ton of as-received coal (Burke et al., 2001; National Petroleum Council, 2007). This range appears to be consistent with the anticipated performance of direct-liquefaction CTL systems under development in Europe and Japan (UK Department of Trade and Industry, 1999; IEA Coal Industry Advisory Board, remarks by Sadao Wasaka, 2006). These yields imply thermal efficiencies of about 62 to 67 percent. Both the yields and the efficiencies apply to direct-liquefaction products that require further upgrading and refining, which will involve additional energy losses, before the products can be used as transportation fuels. Considering energy losses during upgrading and refining, a net process thermal efficiency (CTL

---

[28] The production of this hydrogen is a major source of the carbon dioxide emissions from a direct-liquefaction CTL plant.

plant plus refinery) of roughly 55 to 62 percent appears appropriate, although the thermal efficiency of near-term concepts may not reach the upper portion of this range.

Status of Direct-Liquefaction Development

Friedrich Bergius is credited as the inventor of direct liquefaction, and, often, the overall approach of liquefying coal using heat and pressurized hydrogen is known as the *Bergius process* (Wu and Storch, 1968; Stranges, 2007). Bergius applied for the first patent in 1913. By 1943, German firms were operating 12 direct-liquefaction plants, most of which had been built in the 1930s. German direct-liquefaction CTL production peaked in 1944 at 23 million bpd, more than five times the amount of fuel produced by Germany's FT CTL plants. In the 1930s, a few direct-liquefaction CTL plants were also built outside of Germany, but, with the exception of the Billingham, UK, plant, these plants were unable to successfully operate due to technical problems or were constructed for process development and testing rather than for commercial production. After World War II, a few pilot-scale plants were built, including two in the United States. But abundant petroleum and low oil prices soon caused commercial interest in coal liquefaction to wane.

The 1973 petroleum embargo and price jump revived international interest in using coal to produce liquid fuels. For the most part, this interest coalesced on direct-liquefaction CTL, which was viewed as offering lower costs and higher conversion efficiency than FT CTL. By 1975, the U.S. government's annual budget for direct-liquefaction research and demonstration projects exceeded $100 million and included technical development of more than ten processes for converting coal to liquid fuels (Cohen and Noll, 1991).

By 1980, the U.S. Department of Energy was focusing its efforts on the development of four processes for direct liquefaction: a solvent-refined coal (SRC) process to produce a boiler fuel (SRC-I) and the solvent-refined coal (SRC-II), H-coal, and Exxon donor solvent (EDS) processes to produce liquid fuels suitable for transportation applications (OFE, 1980). Detailed engineering designs were under way for two large demonstration plants (SRC-I and SRC-II), each capable of processing 6,000 tons of coal per day and producing more than 15,000 bpd of synthetic fuels. Two large pilot plants were built and operated to obtain and evaluate scale-up data on the H-coal and EDS processes.[29] By 1981, the annual federal budget for direct liquefaction had grown to $500 million. However, over the subsequent two years, this program was disassembled: The two SRC demonstration plants were canceled, and work at all the major pilot plants was terminated (Cohen and Noll, 1991).

Three factors contributed to the downsizing of federal budgets supporting the development of direct-liquefaction CTL. First, the program was experiencing sig-

---

[29] The EDS pilot plant was designed to receive 250 tons of coal per day. When configured to produce transportation fuels, the H-coal pilot plant could receive 200 tons of coal per day (OFE, 1980).

nificant cost escalation. For example, the initial cost estimate for each of the SRC demonstration plants was $700 million. But as more detailed design information was obtained, these estimates dramatically increased. By the end of 1980, the cost estimates grew to $1.9 billion for the SRC-I plant and $1.4 billion for the SRC-II plant (Cohen and Noll, 1991), and more increases followed. Second, significant technical problems were encountered at the large pilot plants,[30] and there was growing awareness within the U.S. Department of Energy of significant shortfalls in the technology base supporting the design of the two demonstration plants.[31] Third, there was insufficient evidence that direct-liquefaction CTL offered a lower-cost route to transportation fuels than the much lower-risk FT CTL approach.[32]

In spite of greatly reduced federal support, considerable progress was achieved over the following 17 years in improving process yield, selectivity, and product quality (Gorbaty et al., 1989; Burke et al., 2001). In the United States, this work centered on a process variant generally referred to as *two-stage liquefaction*. Process development of two-stage liquefaction was centered on small process-development units located in Wilsonville, Alabama (coal feed of six tons per day), and Lawrenceville, New Jersey (coal feed of three tons per day). Within the United States, the leading proponent of two-stage liquefaction is Headwaters, which has its headquarters in South Jordan, Utah, and CTL research facilities in Lawrenceville, New Jersey.

U.S. government support for direct-liquefaction technology development ended in 1999, most probably in response to the low world oil prices that then prevailed. At present, there is very little activity in direct-liquefaction CTL development outside of China.

China's development effort is currently motivated by plans to develop a domestic CTL industry (Li, 2007). Presently, a 20,000-bpd commercial direct-liquefaction CTL facility is under construction in Majiata, Inner Mongolia, China. This facility is intended to be the first process train for a planned, full-scale direct-liquefaction plant producing more than 100,000 bpd of liquid fuels. The plant is being built by the Shenhua Group, which is a state-owned company heavily invested in coal development. It is our understanding that the Majiata plant incorporates two-stage liquefaction technol-

---

[30] Both the H-coal and EDS pilot plants suffered from major operational problems and equipment failures. Eventually, these problems were overcome, and important design information was obtained.

[31] Specifically, on two occasions in 1980, Harry Perry, then at Resources for the Future, and James T. Bartis, then director of the Office of Plans and Technology Assessment in the U.S. Department of Energy, advised George Fumich, assistant secretary of fossil energy, that the pilot plants supporting the SRC processes did not cover all critical processes and that a 200-fold scale-up (e.g., from 30 tons at the SRC-II pilot plant) was not appropriate for a process involving transfers of solids at high temperatures and pressures.

[32] For example, the only comparative analysis of direct liquefaction and FT CTL then available to the U.S. Department of Energy was a Ralph M. Parsons Company (1979) report that could not make a strong case for direct liquefaction.

ogy, including technology acquired from at least one firm that participated in the U.S. federal program, as well as technology developed by Chinese organizations.[33]

Shenhua's direct-liquefaction facility is scheduled to begin operations in late 2008. Supporting the design of this plant is a six-ton-per-day pilot plant (Li, 2007), which implies a scale-up factor of about 500 for just the first process train. Given this scale-up factor and the fact that the facility incorporates multiple new technologies and involves transfers of solids at high temperatures and pressures, we anticipate that plant start-up will be challenging and could extend well beyond 2008. If successful, the first process train at Shenhua's Majiata plant will be the first large direct-liquefaction plant built since World War II.

Direct-Liquefaction Products

Current information indicates that the technology exists for upgrading and refining gasoline, diesel, and jet fuel produced through direct liquefaction so that they will meet specifications established for the corresponding petroleum-derived products (Gorbaty et al., 1989). In the early 1980s, concerns were raised regarding the mutagenicity, carcinogenicity, and toxicity of liquids produced from direct liquefaction of coal. The raw liquids produced in a direct-liquefaction process are heavily aromatic and include heterocyclic compounds that are known mutagens or carcinogens. This raises a risk of occupational exposure or community exposure through an accidental spill if such raw products were to be transported.[34] If direct-liquefaction CTL is eventually assessed as a viable option, this issue needs to be revisited in light of current standards for worker and public safety.[35]

The extent of upgrading to meet current U.S. or European standards for gasoline and diesel motor fuels will almost surely result in products that pose negligible, if any, risks beyond the corresponding petroleum products. Since current standards for jet fuel require less intensive refining, there may be a residual health risk associated with the handling and transport of jet fuel derived from the direct liquefaction of coal. If products of direct liquefaction are to be used as a jet fuel, appropriate specifications will need to be established (Gorbaty et al., 1989).

---

[33] In its 10-K filing for the year ending September 30, 2006, Headwaters reported the sale of direct-liquefaction CTL technology to the Shenhua Group.

[34] A large body of research was devoted to this issue in the late 1970s and early 1980s (NIOSH, 1981; Wilson et al., 1984; Giddings et al., 1980; and Walsh, Etnier, and Watson, 1981).

[35] If and when the Shenhua plant does successfully operate, this occupational and public-safety issue opens an opportunity for cooperative research between the United States and China.

Baseline Greenhouse-Gas Emissions from Production of Coal-Derived Liquid Fuels

Examining the entire fuel cycle, producing and using diesel fuel and gasoline derived from coal will yield much higher greenhouse-gas emissions than producing and using fuel derived from conventional petroleum. This increase is due to the large amount of carbon dioxide that would be released at an FT CTL plant—about 0.8 tons per barrel of product. Pending the public availability of a conceptual design and engineering analysis for MTG CTL plants, we recommend using the same figure—about 0.8 tons per barrel of product—for MTG CTL plants, because both types of plants operate at roughly the same overall energy efficiency.

There are two causes for these plant-site carbon dioxide emissions. An appreciable fraction of the emissions stems from hydrogen production. The second cause is thermal loss, which occurs because of incomplete recovery of high- and medium-temperature heat and the rejection of low-temperature heat.

For transportation fuels produced in an FT CTL plant without carbon management, we estimate that total-fuel-cycle, mine-to-wheels greenhouse-gas emissions would be 2.0 to 2.2 times above the well-to-wheels emissions associated with fuels produced by refining conventional light crude oils.[36] A significant portion of U.S. transportation fuels are derived from heavier oils, such as those imported from Venezuela and the syncrude produced from Canadian oil sands. For these heavier crudes, the total-fuel-cycle emission ratio (CTL to petroleum-derived fuels) would be appreciably lower. For example, we estimate the total-fuel-cycle greenhouse-gas emission ratio of CTL versus diesel from Canadian tar sands at roughly 1.7 to 1.9.[37]

We specify a range for the total-fuel-cycle ratio because plant-site emissions are sensitive to the overall energy efficiency of the plant. This estimate is for first-of-a-kind moderate-sized or larger plants. Smaller FT CTL plants may exceed this range because their lower overall energy efficiency is slightly lower. Technology advances aimed at significantly improving the energy efficiency and costs of CTL production might eventually be able to reduce plant-site greenhouse-gas emissions by one-fifth.[38] Taking an optimistic view regarding the possibility of such technical progress, producing and

---

[36] See Appendix B. Specifically, our estimate holds for crude oils with an American Petroleum Institute (API) gravity of 35 degrees or higher, such as Arab Light, West Texas Intermediate, Wyoming Sweet, and Brent North Sea crudes. Heavier crudes require more upgrading and refining, which leads to higher greenhouse-gas emissions (Marano and Ciferno, 2002).

[37] This estimate is based on carbon dioxide–emission rates reported in Marano and Ciferno (2002) and Woynillowicz, Severson-Baker, and Raynolds (2005).

[38] Examples of technical advances required to achieve this level of performance include oxygen production at reduced energy consumption, improved gas-gas separation technology, higher-temperature gas-purification systems, and reduced or eliminated oxygen demand during gasification. Such advanced technologies are highly unlikely to be available for FT CTL plants beginning the design process over the next ten years.

using FT or MTG CTL fuel could eventually result in greenhouse-gas emissions that are a factor of 1.6 to 1.8 greater than fuels derived from light crude oils.[39]

For direct liquefaction, the potential 10- to 25-percent increase in total system efficiency[40] implies a 9- to 20-percent reduction in coal use and carbon dioxide emissions as compared to FT CTL using current state-of-the-art technology. For early direct-liquefaction CTL plants, we calculate greenhouse-gas emissions to be a factor of 1.7 to 2.0 greater than those resulting from the production of conventional petroleum–derived fuels. If direct-liquefaction CTL becomes a commercial option, further work on process development might reduce this factor. However, at this time, we see no technical basis for projecting significantly lower greenhouse-gas emissions from a direct-liquefaction plant unless that plant includes provisions for carbon management.

Carbon Capture and Sequestration

The high greenhouse-gas emissions associated with CTL production will likely preclude the development of a large CTL industry in the United States unless plant-site carbon dioxide emissions are managed. In this section, we briefly review the viability of capturing, transporting, and sequestering or otherwise using plant-site emissions of carbon dioxide.

Capturing Carbon Dioxide

The preparation of synthesis gas, or, in the case of direct liquefaction, hydrogen, results in a waste stream of nearly pure carbon dioxide, which would normally be released to the atmosphere.[41] For certain types of FT CTL plants, a second carbon dioxide waste stream is generated just after the FT reactor. In these cases, the technology to separate carbon dioxide from process streams has been commercially available for decades, is well understood, and is reliable. For carbon management, the only additional costs for preparing this waste carbon dioxide for sequestration are the capital and operating costs associated with dehydrating and compressing the waste gas within a range of 2,000–2,200 psi.

The remaining sources of carbon dioxide emissions at a CTL plant stem from the combustion of fuel used for the production of electric power, much of which is required for the cryogenic production of the oxygen used in the gasifiers. A portion or all of this

---

[39] For example, an advanced concept currently under development at Ohio State University offers to increase FT CTL efficiency to 55 percent, thereby decreasing plant-site carbon dioxide emissions by about 20 percent (NETL, 2007g).

[40] From 50 percent for FT or MTG CTL to between 55 and 62 percent for direct-liquefaction CTL plus associated refinery operations.

[41] For further information on carbon dioxide waste streams in CTL plants, see the section "The Fischer-Tropsch CTL Approach" earlier in this chapter.

fuel may be *fuel gases*, which are a by-product of the main liquefaction reactions (i.e., in the FT, MTG, or direct-liquefaction reactors) as well as in product-upgrading steps that occur within the plant.[42] For FT CTL plants, emissions from fuel-gas combustion can represent between 10 and 20 percent of the plant-site carbon dioxide emissions.

An alternative to directly combusting fuel gases is to convert all the hydrocarbon species within the fuel gas to a combination of hydrogen and carbon dioxide and then to remove and capture the carbon dioxide. This conversion of the fuel gases involves processes that are very similar to those currently used in petroleum refineries to produce hydrogen. The carbon dioxide waste stream would be mixed with the other waste streams and compressed. The design of the gas-turbine combustors would need to be modified to allow acceptance of a pure hydrogen fuel.[43]

Overall, the technical risks associated with carbon capture are very low. For FT or MTG CTL plants, the incremental cost involved in capturing between 85 and 90 percent of plant-site emissions is fairly small because the process requires primarily compression of carbon dioxide that has already been separated from the main process streams. With this level of capture, total-fuel-cycle greenhouse-gas emissions from CTL production and use are comparable to emissions from conventional processing of petroleum, assuming that all captured emissions can be successfully sequestered (see Appendix B).

The incremental cost for capturing carbon dioxide associated with fuel-gas consumption at the CTL plant will be higher, but it ought to be comparable to the costs associated with capturing carbon dioxide during the operation of IGCC power-generation systems. With full capture of plant-site carbon dioxide emissions, greenhouse-gas emissions associated with coal-derived liquids should be slightly less (i.e., about 10 percent lower) than emissions associated with conventional-petroleum products. For full capture, we project an impact on product costs of less than $5.00 per barrel.[44]

Transporting Carbon Dioxide

Pipeline transport of carbon dioxide is widely practiced because naturally occurring deposits and synthetic sources of carbon dioxide are used for enhanced oil recovery operations. For example, Kinder Morgan, the largest transporter and marketer of carbon dioxide in the United States, transports more than one billion standard cubic feet of carbon dioxide per day over a 1,300-mile pipeline network (Kinder Morgan,

---

[42] The fuel gas consists of a mixture of nitrogen, hydrogen, carbon monoxide, and light hydrocarbons, such as methane, propane, and butane (NETL, 2007b).

[43] Such modifications are a focus of current U.S. Department of Energy research programs.

[44] As an example, an 88-percent capture of plant-site emissions is expected to cause plant capital expenditures to increase by about 5 percent and electricity sales to drop by 40 kWh per barrel (SSEB, 2006, Appendix D). This would raise CTL production costs by about $4.00 per barrel.

2006).[45] Costs and technical standards for carbon dioxide pipelines are similar to those for long-distance natural-gas pipelines, because they operate at similar pressures. However, carbon dioxide transport is less expensive on a mass basis, because carbon dioxide is considerably denser than natural gas at pressures above 2,000 psi and at ambient temperatures.

Most relevant to prospective CTL plants is the 205-mile pipeline owned and constructed by the Dakota Gasification Company for the purpose of delivering carbon dioxide generated at its coal-gasification plant in Beulah, North Dakota, to oil fields in Weyburn, Saskatchewan. Consisting of 14- and 12-inch-diameter sections, this pipeline is designed to transport 204 million standard cubic feet (11,800 tons) per day of carbon dioxide (Stelter, 2001).

Storing or Disposing of Carbon Dioxide

Every barrel of fuel produced at an FT or MTG CTL plant is accompanied by the production of nearly a ton of carbon dioxide. A very large domestic CTL industry—for example, one producing three million barrels of fuel per day, about one-seventh of current U.S. consumption of liquid fuels—would involve generation at plant sites of about 2.5 million tons of carbon dioxide per day. Mature technology is available to capture this greenhouse gas and to transport it over considerable distances. The technical challenge is how to prevent such a large amount of gas from entering the atmosphere.

Two technical options are currently available for storing the carbon dioxide that would be emitted at CTL plant sites: enhanced oil recovery and geologic sequestration. Each of these options provides a means of reducing total-fuel-cycle greenhouse-gas emissions to levels that are comparable with the use of fuels refined from light crude oils, such as West Texas Intermediate. Besides enhanced oil recovery, no other commercial opportunities currently exist for the disposition of an appreciable fraction of the carbon dioxide emissions associated with strategically significant amounts of CTL production (IPCC, 2005). In the future, such applications may develop. For example, sequestering carbon dioxide in unminable coal seams might promote enhanced recovery of natural gas. Additionally, plant-site carbon dioxide emissions could be captured and used in facilities dedicated to biomass production (e.g., algae farms). Research on such advanced concepts is under way (NETL, 2007a), but, at present, none of these concepts is sufficiently developed to warrant further consideration on the basis of its relevance to the initial stages of the development of a CTL industry.

Enhanced Oil Recovery. Enhanced oil recovery using a method known as *carbon dioxide flooding* is an important carbon-management option, especially for CTL plants located within a few hundred miles of major, active oil basins. Presently, between 30 million and 40 million tons per year of carbon dioxide are used for enhanced oil recovery in the United States, primarily in Texas (Kuuskraa, 2006; IPCC, 2005). Most

---

[45] For carbon dioxide, one billion standard cubic feet per day is about 20 million tons per year.

of this carbon dioxide is being produced from natural reservoirs. An example of carbon dioxide enhanced oil recovery is the Chevron-operated Rangley, Colorado, project, which has undergone carbon dioxide flooding since 1986. In 2003, the carbon dioxide injection at Rangley was 3.3 million tons, of which 0.8 million tons were purchased and 2.5 million tons were recycled from completed flooding operations. Oil production at Rangley has been about 14,000 bpd, or about 1.5 barrels per ton of carbon dioxide injected. Since 1986, net cumulative sequestration has been 24 million tons. Surface release of this cumulative stored gas is below detection limits, implying an annual loss rate of less than 0.0008 percent of total stored carbon dioxide (IPCC, 2005).

An important pioneering effort to establish the sequestration potential of enhanced oil recovery is the Canadian Weyburn project, which combines commercial oil-field operations with an extensive program of monitoring, modeling, and analysis. This is a large project sequestering more than two million tons per year of carbon dioxide.

The 30 million to 40 million tons per year of carbon dioxide currently produced and distributed for enhanced oil recovery in the United States is equivalent to the amount of carbon dioxide that would accompany the production of 100,000 to 130,000 barrels of CTL fuels per day, where CTL fuel production is expressed on a diesel value equivalent (DVE) basis.[46]

Available studies suggest that continued high oil prices and technology trends could make as much as 90 billion barrels of U.S. oil economically recoverable through carbon dioxide flooding (OFE, 2005; NETL, 2007d). A conservative estimate is that this increase in technically recoverable resources will increase U.S. petroleum production by between one million bpd and two million bpd and require the use of between 0.4 million and one million tons per day of carbon dioxide (Kuuskraa, 2006).[47] This amount of carbon dioxide represents production at plant sites associated with a domestic FT CTL industry producing between 0.5 million bpd and 1.2 million bpd.

A benefit of this approach is that oil-field operators currently pay for the carbon dioxide used for enhanced oil recovery operations. For enhanced oil recovery applications, delivered carbon dioxide prices have been reported to be in the range of $10 to $15 per ton (IPCC, 2005). These carbon dioxide prices, however, cover a period (namely, 1990s through 2002) when prices for crude oil were generally between $15 and $23 per barrel (real 2007 dollars). At the much higher crude oil prices that have prevailed since 2005, the demand for carbon dioxide for enhanced oil recovery should be considerably higher, although the long-term market-clearing price will depend on marginal production costs.

---

[46] This calculation is based on CTL plant-site carbon dioxide production of 0.85 tons per barrel of DVE fuel production. DVE fuel production is determined by discounting naphtha production by a factor of 0.71 to account for FT naphtha's lower value relative to FT diesel (see Appendix A).

[47] As historically practiced, each ton of purchased carbon dioxide used in enhanced oil recovery yields, on average, about three barrels of petroleum. In practice, yields range from two to six barrels per ton (IPCC, 2005).

If early CTL plants are able to sell their coproduced carbon dioxide, significant financial benefits could accrue to plant owners. However, uncertainty remains regarding the rate at which carbon dioxide used in various types and locations for enhanced oil recovery will reenter the atmosphere. Resolving this issue is one of the objectives of ongoing domestic and international research efforts (NETL, 2007a).

While carbon dioxide injection for enhanced oil recovery represents an interesting application, it should be viewed as relevant only to the capture of carbon dioxide from the first few CTL plants, especially considering that new coal-fired power plants that capture carbon may also be targeting this market.

Geologic Sequestration. By *geologic sequestration*, we refer to technical approaches being developed in the United States (primarily through funding from the U.S. Department of Energy) and abroad that are directed at the long-term storage of carbon dioxide in various types of geological formations, such as deep saline formations (IPCC, 2005; NETL, 2007a). While the focus and principal motivation of this research is to address carbon dioxide emissions from fossil fuel–fired power plants, successful development of geologic sequestration would also apply to any centralized source of carbon dioxide emissions, including CTL plants.

Geologic carbon sequestration is broadly viewed as the critical technology to allow continued coal use while reducing greenhouse-gas emissions (IPCC, 2005, 2007b; NETL, 2007a; MIT, 2007). There are good reasons to be optimistic that carbon capture and geologic sequestration can be successfully developed as a viable approach for carbon management. The strongest evidence that geologic sequestration can be effective is the existence of naturally occurring deposits of carbon dioxide and other gases that were generated millions of years ago, implying that geologic formations with extremely small leakage rates do exist. Additionally, as of January 2008, three large pioneer carbon sequestration projects are under way: the North Sea Sleipner gas-field project, which injects into a large, subsea saline formation; the aforementioned Weyburn enhanced oil recovery project; and the In Salah project in southern Algeria, which injects into the aqueous layer of a depleted zone of a natural-gas formation. Each of these projects is operating at a carbon dioxide–injection rate of more than one million tons per year, and each is reporting favorable interim results. For example, in the Sleipner project, the fate of the injected carbon dioxide has been successfully monitored via seismic surveys, which have also permitted the development of a refined model that has allowed predictions of the long-term disposition of the stored carbon dioxide. Seismic survey results indicate that the caprock above the saline formation is preventing upward migration of the stored carbon dioxide. Model predictions for the Sleipner project indicate that, over the long term, the carbon dioxide–saturated brine will become denser and sink, eliminating the potential for long-term leakage (see IPCC, 2005, p. 217).

While there are no major technical barriers to geologic sequestration of carbon dioxide, the development of a commercial sequestration capability within the United States requires addressing important knowledge gaps associated with site selection and preparation, predicting long-term retention and leakage rates, and monitoring and modeling the fate of the sequestered carbon dioxide. There are also important legal, regulatory, and public-acceptance issues that must be addressed (NETL, 2007a; IPCC, 2005). Advancing the technology base constitutes a major technical challenge and requires multiple large-scale demonstrations, because geologic conditions can vary considerably among locations. While many have called for demonstrations of sequestration at a scale of one million tons of carbon dioxide per year, this level should be viewed as an intermediate step. For example, the annual plant-site carbon dioxide emissions of a single, moderate-sized CTL plant would be nearly an order of magnitude larger.

Alternative Carbon-Management Options

Other technical approaches to addressing the high greenhouse-gas levels associated with CTL production and use involve either avoiding plant-site emissions or offsetting these emissions by combining CTL production with a biomass-based method of producing liquid fuels.

Eliminating Plant-Site Greenhouse-Gas Emissions

A straightforward approach to eliminating plant-site greenhouse-gas emissions is to deliver sufficient hydrogen to the CTL plant so that all of the carbon in the coal feed to the plant leaves the plant as part of the fuel product. For this to be an effective carbon-management option, the hydrogen sent to the CTL plant must be manufactured without appreciable greenhouse-gas emissions. An example would be hydrogen production via electrolysis using electric power generated from nuclear, wind, or solar energy. This approach has the added benefit of more than doubling the amount of liquid fuel produced per ton of coal delivered to the CTL plant, thereby reducing coal-mining requirements and associated environmental impacts. While eliminating all plant-site greenhouse-gas emissions through hydrogen addition is technically feasible, the high cost of commercial approaches for producing hydrogen from nonfossil sources make this option highly uneconomical in the foreseeable future.

Converting Coal and Biomass to Liquids

Recently, considerable attention has focused on the concept of using a combination of coal and biomass to produce liquids using FT synthesis, including announcements by Rentech and Baard that their planned CTL plants would include gasification of both coal and biomass.

Biomass can be gasified to produce a synthesis gas that can be converted in FT or MTG reactors to fuels that are identical to those that would be produced using coal or natural gas as the gasifier feed.[48] For transportation fuels produced at FT or MTG biomass-to-liquids (BTL) plants, total-fuel-cycle emissions of greenhouse gases are fairly low and are associated primarily with cultivation, harvesting, and delivery of the biomass to the plant. The reason for the low rate of emissions is that all of the carbon in the biomass feedstock was originally obtained from the atmosphere. Carbon dioxide emissions at the plant site or from the combustion of the fuel are balanced by the carbon dioxide absorbed from the atmosphere during the growth of future batches of biomass feed for the plant. Overall, this is a closed cycle except for the small amount of conventional fuel involved in producing the biomass. This is especially the case for non-food-crop biomass, such as corn stover, switchgrass, prairie grass, and forest residues.[49] For these feedstocks, the use and production of FT or MTG liquids results in only 10 to 15 percent of the total fuel-cycle emissions associated with conventional petroleum–based fuels.

To understand the interest in dual-fired coal- and biomass-to-liquids (CBTL) plants, consider a hypothetical case of two side-by-side FT plants, one a CTL plant and one a BTL plant, each producing the same quantity of liquid fuels. The collective greenhouse-gas emission rate for this pair would be the average of the two plants. For example, compared to petroleum-derived fuels, if the total fuel-cycle emission factor is 2.2 for the coal plant and 0.1 for the BTL plant, the emission factor for the pair would be 1.15. To break even with conventional petroleum, the BTL plant in this case would need to produce about 30 percent more fuel than the CTL plant.

However, the two-plant scenario just described has two important flaws. First, practical considerations of the logistics of biomass delivery limit the input of any BTL plant to a few thousand tons per day,[50] implying that a stand-alone BTL plant would likely produce no more than 5,000 barrels of liquid fuels per day. At such production levels, diseconomies of scale cause increased production costs, and smaller equipment sizes generally lead to increased thermal losses and lower overall conversion efficiency. The second flaw concerns the inherent variability associated with obtaining a steady supply of biomass over the operating life of any BTL plant. As is the case with CTL plants, BTL plants are capital intensive, and overall economic viability is very sensitive to any factor that would cause operations to fall below the design capacity of the plant.

---

[48] As defined by the U.S. Department of Energy (undated),

> [T]he term "biomass" means any plant-derived organic matter available on a renewable basis, including dedicated energy crops and trees, agricultural food and feed crops, agricultural crop wastes and residues, wood wastes and residues, aquatic plants, animal wastes, municipal wastes, and other waste materials.

[49] For food crops, conventional-energy consumption is generally higher due to greater use of fertilizers and more intensive cultivation practices.

[50] This holds generally for all BTL plants, whether based on gasification followed by synthesis gas conversion or on the conversion of cellulosic materials, starches, or sugars to alcohols.

Collate Color Text Section

Collate Color Text Section

Technical Viability and Commercial Readiness

CTL production based on FT or MTG synthesis is ready for commercial development in the United States. There is, however, some risk that first-of-a-kind FT or MTG CTL plants could experience severe performance shortfalls or other operational problems, especially during their initial operating years. This risk stems from the limited commercial experience in gasifying U.S. coals and the challenges associated with building a plant that is very large, highly integrated, and technically complex.

With regard to coal gasification, considerable progress in the development of high-temperature, low-methane, low-tar gasifiers has been achieved over the past 25 years.[53] But limited commercial experience in the United States with these gasifiers, as well as with the closely coupled systems for gas cleaning and heat recovery, introduces a certain level of risk. This risk at the coal-gasifier-gas-treatment interfaces is amplified by the overall complexity and magnitude of any CTL project.

Indeed, the sheer size and complexity of an FT CTL plant brings additional risks. The unanticipated difficulties experienced by Sasol in starting up its FT GTL Oryx plant in Qatar are a case in point. A problem with the FT section of the plant seriously constrained initial operational capacity. Postconstruction modifications required to correct this problem caused the plant start-up period, which normally requires a few months, to extend to at least 18 months during which "the Oryx joint venture only generated a marginal cash contribution" (Sasol, 2007a). As Sasol reported, "our experience is that starting up technically complex and first-of-kind facilities takes time and is inherently problematic" (Sasol, 2007b, p. 4).

Because of the complexities associated with coal gasification and the extensive environmental control systems that will be required in any CTL plant built in the United States, FT CTL plants processing U.S. coals will be more technically complex than the Oryx GTL plant. The Sasol experience is not unique; the overall issue of performance shortfalls, process-plant schedule slippage, and start-up difficulties is well known and has been addressed in a series of RAND reports (Merrow, Phillips, and Myers, 1981; Myers et al., 1986).

While FT and MTG CTL are commercially ready, BTL and CBTL are not. Combinations of biomass and coal have been used in commercial plants in the past, but only at low biomass-to-coal ratios and with a limited number of biomass types. The highest ratio used in continuous gasifier operations was at the Nuon IGCC power plant in the Netherlands. Here, the biomass-to-coal ratio (calculated on the basis of energy input) was about one to five. For CTL plants that do not include the capture and sequestration of carbon dioxide, much higher ratios—greater than one to one— would be needed to bring carbon emissions to well-to-wheels parity with petroleum-

---

[53] For FT or MTG CTL plants, the type of gasifiers (dry bottom, moving bed) being used at the Sasol commercial works and also at the Great Plains Synfuels Plant in North Dakota are technically obsolete, unless the plant operators are willing to accept significant amounts of synthetic natural gas (i.e., methane) as a coproduct.

derived fuels. Additionally, the Nuon plant did not use the types of biomass that are estimated to be most abundant in the United States.

Based on prior research directed at FT BTL by the U.S. Department of Energy and the Nuon plant experience, the primary uncertainties or problem areas associated with CBTL involve establishing viable approaches to preparing and injecting biomass into the gasifier and the subsequent effect that the biomass ash may have on the internal walls of the gasifier (Ratafia-Brown, 2007). Neither of these issues is a serious scientific challenge, but they must both be resolved before full-scale units are built. Technical resolution is complicated by the inherent variability of biomass and biomass ash. At a minimum, a solid technical basis for a commercial CBTL plant is five years into the future, because demonstration of commercial viability will require the construction and operation of a moderately large test facility or modifications to and testing of gasifiers operating at an existing commercial facility. Pending completion of this development and testing work, CBTL plants will be limited to accepting a fairly low biomass fraction—most likely 20 percent or less on an energy-input basis.

Direct liquefaction will not be commercially ready until performance is established by Shenhua or through the construction and successful operation of a large, integrated pilot plant that processes hundreds of tons of coal per day.

Production Costs

A major uncertainty associated with CTL development is the anticipated cost of producing liquid fuels. For FT CTL costs, the best publicly available studies are based on work sponsored by the National Energy Technology Laboratory—namely, work performed by David Gray and associates at Noblis (formerly Mitretek) (SSEB, 2006); Bechtel and Amoco (Bechtel, 1998); and National Energy Technology Laboratory engineering staff (NETL, 2007b). The cost information provided by these studies is based on low-definition, conceptual designs that were executed for the purposes of guiding federal research and development (R&D) efforts and understanding potential economic viability. In particular, none of the publicly available designs is site specific, and none includes the level of engineering detail required for reliable costing and investment decisionmaking. Prior RAND research on design definition and costing indicates that cost growth is likely to occur as more definitive design work is completed (Merrow, Phillips, and Myers, 1981). For this reason, we derive from the previously published engineering studies an estimate of production costs, particularly capital costs, that is considerably higher.

For MTG CTL, publicly available engineering and cost analyses are more than 20 years old (Wham et al., 1981; Irvine et al., 1984) and do not incorporate recent advances in process technology, especially in coal gasification and the lessons learned from the MTG GTL–operating experience in New Zealand. MTG CTL plants, however, are very similar to FT CTL plants, and the two types of CTL plants should have similar capital costs and operating costs. Which technology—FT or MTG—makes most sense for a particular project may depend more on project and site-specific details and the regional market for the plant products (middle distillates and naphtha from an FT CTL plant versus gasoline and LPG from an MTG CTL plant) than on any absolute cost advantage of one over the other.

CTL plants are capital intensive. For moderate to large FT CTL plants producing diesel fuel, naphtha, and electricity, we estimate capital investment costs of $100,000 to $125,000 (January 2007 dollars) per barrel of daily production capacity, on a DVE basis. These are overnight construction costs and do not include interest accumulated during construction (see Appendix A for details). The low end of our estimate (our reference-case estimate) is considerably higher than most published estimates.[54] This estimate incorporates a 25-percent project contingency,[55] which we judge to be minimal, given the level of design definition that underlies publicly available cost estimates and the lack of recent experience in designing and constructing FT CTL plants. The high end of our capital cost estimate, which raises overall investment costs by an additional 25 percent, is based on prior RAND research[56] as well as on recent reports of escalating construction costs in the process industries.[57] This range is in agreement with a Sasol cost estimate of $100,000 to $125,000 per barrel of daily production capacity for a large (80,000 bpd) plant built in the United States (Roets, 2006).

We estimate nonfuel operating costs at between $7 and $12 per barrel (DVE), assuming a credit for electricity sales. The lower bound of this estimate, which we use as our reference case, is consistent with recent engineering analyses of CTL plants processing bituminous coals (SSEB, 2006, Appendix D; NETL, 2007b). The upper bound, which is used in our high-cost case in Chapter Seven, represents the impact of raising fixed operating costs by one-third (see Appendix A).

With these ranges for capital and operating costs, we estimate that FT diesel can be produced at $1.70 to $2.00 per gallon (January 2007 dollars). This estimate is directly comparable to refinery gate prices for ultralow-sulfur diesel (ULSD).[58] For FT CTL fuels to be competitive, the required selling price for crude oil, using West Texas

---

[54] This estimate derives from Gray's case 3 design (30,000 bpd plant with recycle; see SSEB, 2006, Appendix D) by escalating to January 2007 dollars and replacing Gray's 5-percent contingency with a 25-percent contingency.

[55] Suggested by David Gray (2007) as a reasonable contingency for first-of-a-kind plants.

[56] RAND research conducted during the 1980s found that, for an appreciable fraction of pioneer process plants, cost increases of 50 percent were common for estimates performed during project definition or early stages of engineering (Merrow, Phillips, and Myers, 1981).

[57] For example, in 2006, a National Energy Board of Canada cost estimate for tar sand development showed a 50-percent increase over estimates made by the same board two years earlier (National Energy Board, 2006).

[58] For comparison, the average wholesale price of ultralow-sulfur diesel fuel during the last six months of 2007 was $2.41 per gallon (EIA, 2008d).

Intermediate as the benchmark, must be between $55 (reference case) and $65 (high-cost case) per barrel. Again, we emphasize that all publicly available information on CTL plant costs is based on low-definition designs. Until far more detailed, site-specific design information is available, the costs of CTL production remain highly uncertain and could fall outside of the $55 to $65 per barrel crude oil equivalent range.

These cost estimates are for first-of-a-kind FT CTL plants operating on bituminous coal and producing at least 25,000 barrels of liquids per day. Production costs for smaller plants will likely be considerably higher. Our cost estimates are based on the financial assumptions described in Appendix A. Specifically, we assume 100-percent equity financing; a moderate, real, after-tax rate of return of 10 percent,[59] which is above the U.S. average for corporate returns on investments; and a delivered coal price of $30 per ton.[60]

These CTL production-cost estimates include the extra costs associated with capturing carbon dioxide and preparing it for pipeline transport, under the assumption that the CTL plants will be able to find a potential user who will take the captured carbon dioxide at no cost to the CTL plant. Early CTL plants may be able to sell captured carbon dioxide and thereby generate income that can offset a portion of the costs attributable to CTL production. For CTL plants that are required to pay for the transport and sequestration of carbon dioxide, we estimate that production costs will increase by about $7 per barrel.[61]

Estimates of production costs are highly sensitive to the after-tax real rate of return anticipated by investors in a CTL project. Investors requiring a higher rate of return will not invest in a CTL plant unless they believe that crude oil prices will be even higher than the $55 to $65 range. Illustrating this effect for a 100-percent equity-financed CTL plant, Figure 3.5 shows, for example, that expectation of a 12-percent after-tax real rate of return requires average crude oil prices to be between $62 and $75 per barrel. Lower production costs might be possible through debt financing. The impact of debt financing is discussed in Chapter Seven and in the companion volume to this book (Camm, Bartis, and Bushman, 2008).

Because subbituminous coal is less expensive, overall production costs of FT CTL plants located in Wyoming and Montana may be lower by as much as $5.00 per

---

[59] The 10-percent real after-tax rate of return corresponds to a 20-percent nominal pretax rate of return, assuming an annual inflation rate of 2.5 percent and the federal and state tax rates listed in Appendix A.

[60] For each $1.00 per ton increase in delivered coal prices, the required crude oil selling price increases by $0.55 (see Appendix A).

[61] The $7 per barrel estimate assumes 0.7 ton of captured carbon dioxide per barrel of liquid production, pipeline costs of $4 per ton, and sequestration and monitoring costs of $6 per ton of captured carbon dioxide. Both cost estimates are on the higher side of estimates published by the Intergovernmental Panel on Climate Change (2005).

Figure 3.5
Estimated Required Crude Oil Selling Price Versus Rate of Return for 100-Percent Equity-Financed Coal-to-Liquids Plants

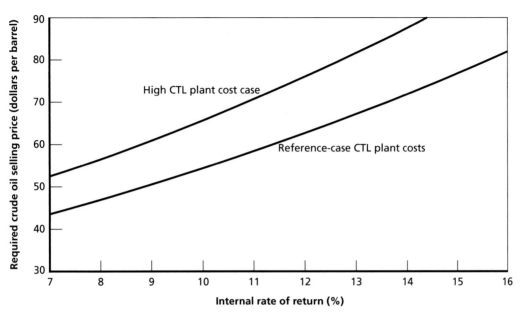

barrel.[62] For lignite, we anticipate that required increases in capital and operational costs will offset the cost advantage that some lignites have over bituminous coals.

Our cost estimates apply only to the first generation of FT CTL plants built in the United States. The cost of building and operating new plants will drop significantly once early commercial plants begin production and experience-based learning is under way. Earlier RAND studies suggested that FT CTL plants might achieve substantial cost reductions, perhaps as much as a 30-percent decrease in production costs per doubling of cumulative production (Merrow, 1989).[63] If FT CTL becomes an economically viable option, it is reasonable to anticipate that, within the next decade, global production levels would rise from the current 140,000 bpd (by Sasol) to at least 500,000 bpd. At that level of production and with a 30-percent learning factor, our estimated production cost range would drop to $38 to $46 per barrel by 2020.[64]

---

[62] Gray obtained a decrease of nearly $10 per barrel, but at least half of this difference is due to the use of more efficient technology in the subbituminous case (SSEB, 2006, Appendix D).

[63] Learning is not guaranteed. It will depend on management attention to R&D, effective information transfer, and organizational capabilities and continuity. There is also considerable uncertainty in the projected rate of learning. For example, Merrow's (1989) estimated cost-improvement factor ranges from 23 to 36 percent.

[64] This estimate is based on the assumption that cumulative global CTL production will approach six billion barrels by 2018, which is roughly a doubling of cumulative production to date in South Africa.

While the preceding cost estimates were derived from engineering analyses of FT CTL plants, they also constitute the best available estimates for MTG CTL plants, pending the public availability of an engineering analysis of an MTG CTL plant.

Timeline for Coal-to-Liquids Development

To address the issue of how fast a CTL industry might develop in the United States, we assume that crude oil prices remain high, that further progress in commercial development proves that CTL is a competitive alternative, and that investment barriers impeding the initial round of CTL plants are removed. Under those conditions, how much CTL production can we expect?

Our research shows that the controlling factors for the growth of a CTL industry in the United States are economic considerations (beyond competitiveness with crude oil) and resolution of the technical, regulatory, and legal issues associated with controlling greenhouse-gas emissions. Ignoring these two factors, a CTL industry can be developed very quickly. One example is the aggressive schedule published by the Southern States Energy Board—namely, one million barrels of CTL by 2016, two million by 2019, and 5.6 million by 2030 (SSEB, 2006).

Economic considerations suggest that it would be prudent to adopt a considerably slower pace to capture the benefits of experience-based learning and to avoid escalation of construction costs (i.e., cost-factor inflation) associated with a rapid ramp-up of demand for specialized design and construction services, manufactured process components, and construction materials.

To capture experience-based learning, we anticipate that the first stage of CTL development in the United States would be limited to a few plants for the purpose of gaining early commercial experience. Once this early experience is obtained, CTL development would transition to a production-growth stage in which multiple plants could be built each year. Assuming that decisions to build an initial round of CTL plants are made in 2008 and that six years are required to build and gain initial operating experience on this first round of plants, the first set of plants in the production-growth phase would not enter their final design stage until 2014 and would not begin commercial operations until 2018 at the earliest.

The relationship between cost escalation and potential ramp-up rates of a CTL industry is difficult to quantify because it is highly dependent on the supply and demand for services and materials associated not only with CTL development but also with other construction and economic-development activities. As a rough estimate, we assume that the first year of the production-growth phase is limited to 150,000 bpd of new capacity and that annual capacity additions increase by 10 percent per year until they reach a maximum of 300,000 bpd.

Including considerations for both learning and cost escalation, we obtain the maximum annual production rates shown in Table 3.1 in the row titled "Economic constraints." For purposes of comparison, we have included an unconstrained case derived from the Southern States Energy Board (2006).[65]

Any national program that controls greenhouse-gas emissions will likely involve imposing a cost on the release of carbon dioxide into the atmosphere. If these costs are sufficiently high to prohibit additional coal use without carbon capture and sequestration, CTL development in the United States will not enter the production-growth stage until the technical, economic, and environmental viability of large-scale carbon sequestration is demonstrated. With an accelerated schedule for multiple, large-scale demonstrations of geologic sequestration, it may be possible to establish carbon capture and sequestration viability for at least a few geologies and technical approaches by 2015. This schedule would allow the first set of plants in the CTL production-growth phase to begin commercial operations in 2020. The out-year production implications of this two-year delay, relative to the economic-constraint case, are shown in Table 3.1.

In summary, for the United States, our analyses indicate that economic constraints and the time required to bring carbon capture and sequestration to commercial viability will limit the maximum rate of CTL industrial development. By 2020, the maximum production level would be about 500,000 bpd. Post-2020 capacity buildup could be fairly rapid, with U.S.-based CTL production in the range of three million bpd by 2030.

While the United States leads the world in proven coal reserves, the next five nations[66] together contain roughly 540 billion tons of proven coal reserves (Table 1.1 in Chapter One), which is about twice the holdings of the United States. Of these nations, China is actively pursuing early commercial experience in CTL and already has under way a number of commercial projects for producing chemicals from coal gasification. Because of this early experience, buildup of CTL production

Table 3.1
Coal-to-Liquids Development Timelines Showing Constraints That Reduce Estimated Maximum Coal-to-Liquids Production Levels (millions bpd)

Constraint	2015	2020	2025	2030	2035
Unconstrained case	0.6	2.2	4.1	5.4	>6
Economic constraints	0.2	0.7	2.0	3.5	5
Accelerated carbon capture and sequestration demonstration	0.2	0.5	1.5	2.9	4.4

---

[65] The unconstrained case shown in Table 3.1 is identical to the Southern States Energy Board's CTL production estimates, moved back by one year to account for the slow progress in CTL industrial development since 2005.

[66] Russia, China, India, Australia, and South Africa.

capacity in China could occur more quickly than in the United States, though other considerations, including increasing demand for coal to support electric-power generation, may limit CTL production growth in China.

Of the remaining four large coal-holding nations, none has extensive commercial experience in gasifying its domestic coals using modern technology. If the uncertainties associated with CTL development are favorably resolved, the rate of building CTL production capacity in these nations will likely be limited by the same constraints discussed earlier for the United States.

Just as prevailing uncertainties prevent a prediction of how much CTL capacity will be built in the United States over the next few decades, an analytical basis for predicting global CTL production growth is not available. But considering the diversity of countries holding large coal resources, we see no reason to anticipate that maximum annual CTL production levels outside the United States should not exceed those of the United States, as shown in the bottom row of Table 3.1.

CHAPTER FOUR

Other Unconventional Fuels

Presently, petroleum demand in the United States stands at between 20 million and 21 million bpd. Imports meet 60 percent of this demand: ten million bpd of imported crude oil and 2.5 million bpd of imported petroleum products. Analyses of U.S. energy requirements for the next 20 to 25 years generally show slightly growing demand for liquid fuels and continued high dependence on imported petroleum. Looking at global trends, we anticipate that rapidly growing energy demand from large developing nations, such as China and India, will raise global petroleum consumption by 20 to 50 percent beyond current levels by 2030.[1] These trends strongly suggest that, without significant additional changes in the energy policies of the United States, we should anticipate petroleum imports in the next few decades to range between 10 million and 12 million bpd (EIA, 2008c). Moreover, unless nations with large or rapidly growing economies make significant changes to their energy policies, we should anticipate growing world demand for petroleum, a long-term trend toward higher prices, and increased dependence on supplies from the Middle East.

In this chapter, we review other approaches for using domestic resources to produce transportation fuels that can substitute for conventional petroleum and lessen U.S. dependence on imports. Our focus is on understanding whether and to what extent other unconventional-fuel options are available to produce liquid fuels that can substitute for conventional petroleum–derived products.[2]

We excluded certain domestic petroleum resources that might qualify as unconventional but were beyond the scope of our study. These include heavy oils and oil deposits that require advanced enhanced oil recovery methods. We also excluded consideration of long-range, high-risk concepts (such as using hydrogen as a transportation fuel and all-electric cars) and advanced biomass concepts (such as genetically engi-

---

[1] Current world demand for liquid fuels is about 85 million bpd (including crude and natural-gas plant liquids) (EIA, 2008b, Table 1.7). Examples of projected 2030 petroleum demand are the International Energy Agency's reference-case projection—116 million bpd (IEA, 2007)—and EIA's 2008 reference- and high-oil-price-case projections, 113 million and 98 million bpd (EIA, 2008c, Table C6).

[2] Heavy government promotion of certain of these fuels has raised important public policy issues, but we do not address these issues here.

neered algae). While research on these approaches may be worthy of support, insufficient information is available to speculate on what, if any, contribution to the U.S. fuel supply they offer for the next few decades.

The more we learned about the benefits of reducing dependence on conventional sources of petroleum, the more we appreciated the importance of improved energy conservation and energy efficiency as ways to accomplish this goal. Understanding the opportunities in these areas, as well as exploring such interesting concepts as plug-in hybrid vehicles, is extremely important but beyond the scope of our study.

Commercially Ready Unconventional Fuels

In addition to FT CTL and MTG CTL, two other approaches are commercially available for producing liquid fuels that can substitute for petroleum: fermenting food crops to produce alcohols and deriving fuels from renewable oils, such as soybean oil.

Food-Crop Alcohols

In the United States, fuel ethanol is fermented primarily from corn and is used as a blendstock for gasoline. Production in 2007 averaged 423,000 bpd (EIA, 2008a), which is the energy equivalent of about 280,000 bpd of gasoline. This production is driven by federal requirements to include oxygenates in automotive gasoline and by a federal subsidy of $0.51 per gallon of ethanol, which corresponds to $0.76 for the amount of energy in a gallon of gasoline or about $35 for the amount of energy in one barrel of crude oil.

The Energy Independence and Security Act of 2007 (P.L. 110-140) extended and increased the renewable-fuel standard originally set in the Energy Policy Act of 2005 (P.L. 109-58). The law now requires a minimum of nine billion gallons of ethanol in transportation fuels used in the United States during 2008. This is the energy equivalent of roughly 400,000 bpd of gasoline. For ethanol fermented from corn, this mandate rises to 13.2 billion gallons per year by 2012, the energy equivalent of about 590,000 bpd of gasoline, which is about 4 percent of the projected demand for transportation fuels in that time frame (EIA, 2008c).

Market demands for corn as food will limit use of corn grain to produce ethanol. For example, based on U.S. Department of Agriculture (USDA) projections (Interagency Agricultural Projections Committee, 2007), we estimate that about one-third of the U.S. corn harvest in 2012 will be devoted to producing ethanol. This would more than double the 14 percent used in the 2005–2006 harvest year and would take an area about 85 percent of the size of Illinois to grow enough corn for this quantity of ethanol. For 2030, EIA's projection for corn-based fuel alcohol production is 660,000 bpd (gasoline energy equivalent),[3] still a small fraction of the total projected demand for transportation fuels (EIA, 2008c).

In 2006, DuPont and BP announced that they were forming a partnership to produce and market butanol (see DuPont, 2007). This effort centers on a demonstration facility in the United Kingdom that produces butanol from sugar beets. The plant is scheduled to begin operations in early 2009. In late 2007, the DuPont/BP partnership plans to deliver "market development quantities" of biobutanol to the United Kingdom (DuPont, 2007).

Butanol is far superior to ethanol as an automotive fuel. It has an energy density much closer to that of gasoline. It offers superior environmental performance because its vapor pressure is lower than ethanol's. It can be blended with gasoline at refineries, thereby eliminating the need for the special handling required by ethanol. It does not require modifications to automobile engines, even when used in high-butanol blends. These factors give butanol a significant competitive edge over ethanol. If butanol fermentation technology progresses to the stage at which it is competitive with ethanol fermentation, demand for alcohol fuels produced from food crops should increase because of butanol's superior properties. Nonetheless, the lack of suitable acreage for additional crop production to satisfy demands for both food and fuel alcohol will likely limit food-based alcohol fuel production (ethanol and butanol) in the United States to less than one million bpd, gasoline energy equivalent.

Fuels Derived from Renewable Oils

To date, vegetable-oil-based fuels have seen limited application in the United States. The dominant U.S. feedstock for the production of biodiesel fuel is soybean oil. Other feedstocks for biodiesel include sunflower oil, rapeseed (canola) oil, beef tallow and other animal fats, and waste cooking grease. In 2007, the production and consumption of biodiesel in the United States averaged 32,000 bpd (EIA, 2008c, Table 10.3), which is about 0.02 percent of U.S. petroleum demand for transportation.

The renewable-fuel standard established by the Energy Independence and Security Act of 2007 (P.L. 110-140) mandates that, by 2012, biomass-based diesel use will average about 65,000 bpd, about twice the estimated biodiesel use in 2007. The costs and market impacts of this mandate, including impacts on food costs, remain highly uncertain.

The resource base of soybeans as a feedstock for biofuel production is limited. Soybeans are one of the major crops grown in the United States. Dedicating the equivalent of the entire soybean crop to biodiesel production would yield 296,000 bpd of fuel.[4] This small amount of production would require the cultivation of a very large

---

[3] This projection is for the EIA reference case and assumes continued federal subsidies.

[4] This result is based on the following data: Total 2005 soybean production was 3.06 billion bushels grown on 71.4 million acres. Biodiesel-fuel yield from soybeans is estimated at 3.8 barrels per 100 bushels. Average soybean

amount of land, an area just slightly smaller than the states of Illinois and Iowa combined. Because soybeans are often planted in rotation with corn, increased corn plantings are expected to reduce the acreage available for soybeans. For these reasons, the World Agricultural Outlook Board of USDA estimated, prior to the passage of the Energy Independence and Security Act (P.L. 110-140), that biodiesel production will level off at approximately 700 million gallons per year (46,000 bpd), using 23 percent of soybean-oil production but displacing only 1.4 percent of the projected demand for diesel fuel (Interagency Agricultural Projections Committee, 2007; EIA, 2008c). In the foreseeable future, vegetable oils will likely provide no more than a few tens of thousands of barrels of fuel per day, considering the importance of soybeans to human food supplies, production costs, environmental impacts (Hill et al., 2006), and fuel suitability.

A small amount of additional production can be achieved from animal fats. For example, in 2007, Tyson Foods announced plans for two ventures that, if fully exploited, might eventually produce about 16,000 bpd of diesel fuel.[5] Based on these statements, it is reasonable to assume that a few tens of thousands of barrels of fuel per day might eventually be produced by processing animal fats, depending on future prices for diesel fuel and animal fat and continuing government subsidies and support.

Emerging Unconventional Fuels

Oil Shale

The largest and richest oil shale deposits in the world are located in the Green River Formation, which covers portions of Colorado, Utah, and Wyoming. Total potentially recoverable resources are estimated at roughly 800 billion barrels, which is more than triple the total petroleum reserves of Saudi Arabia. This is also significantly more than the amount of liquid fuel that could be produced from the proven recoverable coal reserves of the United States, especially considering the role of coal in generating electric power.

In a recent publication (Bartis et al., 2005), RAND researchers reviewed the prospects and policy issues associated with oil shale development in the United States. That book concluded that the prospects are uncertain. They depend primarily on the successful development of in-situ methods that offer improved economics and reduced adverse environmental impacts, as compared to mining the oil shale and retorting it above ground. Since publication of that book in 2005, important technical progress

---

yield in 2005–2006 was 43.0 bushels per acre (Interagency Agricultural Projections Committee, 2007).

[5] This estimate is based on 175 million gallons per year through an alliance with ConocoPhillips (Tyson Foods, 2007a) and 75 million gallons per year from the initial facility built under a joint venture with Syntroleum Corporation (Tyson Foods, 2007b).

has taken place. A number of highly qualified firms have either publicly announced or indicated to us their interest in developing oil shale. In December 2006, the Bureau of Land Management (BLM) announced that it was issuing to three firms—Shell, Chevron, and EGL Resources—small lease tracts in Colorado for the purposes of conducting research, development, and demonstration (RD&D) of in-situ methods for producing fuels from oil shale. In April 2007, BLM announced an additional RD&D lease in Utah to a fourth firm—Oil Shale Exploration Company—that is interested in developing a method involving mining and surface retorting. Other firms have expressed interest in participating in the BLM program if a second round of RD&D leases becomes available.

Based on our knowledge of firms interested in oil shale development, none—with the possible exception of Shell—will be prepared to make a financial commitment to a pioneer commercial-scale facility for at least five and, in some cases, as many as ten years. Accordingly, we do not anticipate that commercial production will exceed 100,000 to 200,000 bpd for at least the next 15 years.

It is possible that oil shale might ultimately make a significant contribution to the U.S. transportation-fuel supply. However, a production level of one million bpd is probably more than 20 years in the future, and three million bpd is probably more than 30 years in the future (Bartis et al., 2005).

U.S. Tar Sands

The most recent information on U.S. tar sand resources is contained in USGS reports that summarize resource estimates made in the 1970s, 1980s, and 1990s (Meyer, Attanasi, and Freeman, 2007; USGS, 2006) and a report by the University of Utah prepared for the U.S. Department of Energy Office of Fossil Energy and the National Energy Technology Laboratory (Utah Heavy Oil Program, 2007). Major U.S. bitumen deposits (more than 100 million barrels) can be found in Alabama, Alaska, California, Kentucky, New Mexico, Oklahoma, Texas, Utah, and Wyoming.[6] Measured resources of natural bitumen in place are 36 billion barrels (USGS, 2006). There are an additional 18 billion to 40 billion barrels of speculative resources in place.[7] The largest and best-defined tar sand deposits in the United States are in Utah, which holds about 20 billion barrels (measured and speculative) in large deposits. The estimates of tar sand resources in place do not address how much of that resource may be recoverable and at what cost.

Exploitation of U.S. tar sands would conceivably use a combination of surface mining and in-situ extraction methods. Such methods are being applied to extract bitumen from the more abundant and richer Canadian tar sands (often referred to

---

6   Bitumen is the very viscous organic liquid found in tar sands.

7   Speculative resource estimates are highly uncertain because they are typically the result of extrapolating observations of surface and near-surface resources to resources that might be located deeper underground.

as oil sands) located in Alberta. However, U.S. tar sands are hydrocarbon wetted and often contained in solid rock, whereas the more abundant and richer Canadian tar sands located in Alberta are water wetted and contained in loose sand (Task Force on Strategic Unconventional Fuels, 2007).[8] For these and other reasons (Utah Heavy Oil Program, 2007), extraction techniques for U.S. tar sands are likely to be different and more costly than those used in Alberta.

Most of the interest in developing U.S. tar sand resources appears to emanate from Washington, D.C. For example, Section 369 of the Energy Policy Act of 2005 (P.L. 109-58) directs the U.S. Department of Energy to make public lands available for lease to promote R&D and to prepare a programmatic environmental-impact statement regarding the production of liquid fuels from oil shale and tar sands. Because of the extensive oil-sand development occurring in Alberta, numerous firms have experience in producing liquid fuels from tar sands. Of this group, or among firms of similar capabilities, none has announced interest in U.S. tar sand development.[9]

While small-scale development of U.S. tar sand resources might occur for other purposes, such as materials for road surfacing (Foy, 2006), we conclude that, in the foreseeable future, U.S. tar sand deposits are not likely to be a viable source for the production of significant quantities of liquid fuels.

Biomass from Nonfood Crops

Two approaches are potentially available for producing liquid fuels from renewable biomass resources other than food crops: alcohol production via cellulosic fermentation and gasification followed by FT or MTG synthesis. With either approach, between 0.6 and 0.7 tons of biomass should yield one barrel of liquid fuels with roughly the energy equivalent of one barrel of crude oil. By expanding the biomass resource base beyond food crops, these two approaches substantially increase the potential liquid-fuel production level beyond one million bpd, the upper bound for fuels derived from food crops.

Appropriate biomass resources for cellulosic fermentation or gasification include dedicated energy crops that would be cultivated specifically for the purpose of producing liquid fuels. Examples include switchgrass and poplar bred to maximize mass yield. Recent attention has been directed at the environmental benefits of cultivating mixed prairie grasses, which are mixtures of indigenous grasses that require few agricultural inputs (Tilman, Hill, and Lehman, 2006). The second broad category of biomass resources is residues. Included in this category are agricultural residues, such as

---

[8] In hydrocarbon-wetted sands, the bitumen is in direct contact with the sand grains. In water-wetted sands, a thin layer of water separates the bitumen from the sand grains. Steam-based methods, such as those employed in Alberta, are generally less effective when applied to hydrocarbon-wetted sands.

[9] A recent report promoting the development of oil shale and tar sands identified only six firms interested in tar sands, none of which had experience in building and operating commercial processing systems similar to those that would be associated with tar-sand development (DOE, 2007b).

corn stover, wheat straw, and manure; limbs and other tree parts left over from logging operations; forest thinnings; and municipal wastes, including municipal wood waste and yard trimmings, municipal solid wastes, and sewage sludge.

Several recent analyses have investigated the potential biomass resource that exists in the United States. A study by Oak Ridge National Laboratory (Perlack et al., 2005) estimated that approximately 330 million dry tons per year of forest residue and thinnings, urban wood waste, and agricultural residue are available annually, assuming no changes in land use or increases in agricultural yields. Assuming modest changes in agricultural yields and cultivation of energy crops on idle land, the National Renewable Energy Laboratory (Milbrandt, 2005) performed a similar analysis and estimated that 360 million dry tons per year of such biomass resources exist. The Oak Ridge estimate increases to about 1.4 billion tons annually under the following conditions: more-intensive collection efforts, significantly improved agricultural yields, land-use changes, and inclusion of perennial bioenergy crops.[10] Among estimates supported by analytic methodology, the Oak Ridge estimate forecasts the highest potential tonnage of biomass resources for energy use. This estimate is based solely on land availability and does not consider the cost of collecting and delivering biomass to sites where it would be used to produce electric power, useful chemicals, or liquid fuels. Assuming that half of this biomass can be economically collected and delivered to liquid fuel–production facilities, biomass resources would be able to support a fuel-production level of roughly three million bpd.

As compared to fossil energy resources, biomass must be collected from large areas. Annual yields of biomass crops range from two to ten dry tons per acre. An FT BTL plant producing 4,000 barrels of liquids per day would require delivery of between 800,000 and 900,000 dry tons per year. That level of demand would require dedicated cultivation of between 100,000 and 400,000 acres. Taking into account the density of land available for biomass production, along with field and storage losses, the land area over which a single plant would be supported could exceed one million acres, which is roughly the area contained in a circle with a radius of 25 miles.

Alcohol Fuels from Cellulosic Feedstocks. The sugars and starches that are often the focus of food-crop cultivation are generally a relatively small fraction of the weight of a plant. Most of the material in plants is cellulose, hemicellulose, or lignin. None of these substances is amenable to the fermentation procedures that are commonly used to produce ethanol. To overcome this problem, researchers have devised approaches that break down the cellulose and hemicellulose into sugars. These sugars would then be converted to ethanol, and possibly butanol, through fermentation.

Most of the federal effort to derive liquids from cellulosic feedstocks centers on producing ethanol via fermentation. Primary emphasis is on using enzymes to break

---

[10] Under this estimate, perennial bioenergy crops would be grown on about 55 million acres of cropland, idle cropland, and cropland pasture.

down the cellulosic materials into the simple sugar glucose. In February 2007, the U.S. Department of Energy announced federal funding of up to $385 million for commercial demonstration of six initial cellulosic-ethanol plants (DOE, 2007a), five of which involve fermentation.[11] The six demonstration facilities are to be constructed between 2007 and 2011, and their target capacities range from 440 to 2,040 bpd of ethanol. The processes underlying the proposed facilities have been demonstrated at the pilot scale (i.e., on the order of 1/100th of the proposed capacities).

The cost of producing ethanol fuels from these initial plants will be very expensive, but this should be expected. After all, they are demonstration facilities operating at scales well beyond prior experience. At present, insufficient information is available to allow us to predict the costs of producing ethanol from commercial facilities that might be built using the information obtained during the ongoing demonstrations. Several years of operating experience of these and other facilities should eventually resolve uncertainties regarding the technical, economic, and environmental viability of the various conversion processes.

Biomass Gasification and Liquids Synthesis. A second approach for converting cellulosic biomass into liquid fuels is to gasify the biomass, produce synthesis gas, and, after cleaning, convert that gas to liquids in an FT reactor (Boerrigter and van der Drift, 2004; Boerrigter and Zwart, 2005) or via the MTG process. Most of the steps in this approach are identical to those that would be encountered in a CTL plant or a natural-gas-to-liquids plant. For this reason, the FT and MTG BTL approaches are of much lower technical risk than all approaches involving alcohol production from cellulosic materials via fermentation. We have not found convincing evidence that either process—gasification and liquids synthesis or fermentation—offers economic advantages over the other.

For the FT and MTG BTL approaches, the principal development need is for improved technology for handling, preparing, and feeding biomass into a pressurized gasifier.

The use of a combination of biomass and coal may be an approach that significantly lowers the costs and investment risks involved in building plants that use biomass to produce liquid transportation fuels. First, biomass yields are very sensitive to weather, especially rainfall. Cofiring a gasifier with both biomass and coal offers a means of smoothing out this inherent fluctuation in biomass availability. Second, plants that receive only biomass are limited in size. Otherwise, the land area supporting the plant becomes very large, delivery of biomass to the plant becomes expensive, and the amount of truck traffic required to deliver the biomass to the plant may be viewed as unacceptable. Cofiring with both biomass and coal offers a way to build larger plants that can capture economies of scale while avoiding the problems of very

---

[11] One of the selected projects involves gasification of biomass and conversion of the resulting synthesis gas to mixed alcohols.

large biomass-collection areas. Third, combining coal-biomass cofiring with capture and sequestration of carbon dioxide offers a way to reduce greenhouse-gas emissions to well below those associated with conventional-petroleum fuels, as discussed in Chapter Three.

Missing from the federal R&D portfolio are any research efforts to establish the commercial viability of a few techniques for the combined use of coal and biomass. The most pressing near-term research requirement centers on developing an integrated fuel-processing and gasification system capable of handling both biomass and coal, as discussed in Chapter Three.

Summary

Looking ahead to the next 30 to 50 years, there is no single unconventional-fuel technology capable of meeting U.S. demand for liquid transportation fuels and continued dependence on imported crude oil. At present, FT and MTG are two of only three unconventional-fuel technologies that are commercially ready and capable of producing significant amounts of fuel. The other is food-crop fermentation to produce ethanol. However, the ultimate potential of ethanol derived from food crops is less than one million bpd.

Biomass and oil shale, once commercial, might together be capable of supplying up to six million bpd, but building to this level of production would likely require 30 years. Coal offers the possibility of producing almost three million bpd by 2030. Most importantly, FT and MTG CTL technologies are ready for initial commercial applications. Moreover, FT and MTG CTL may also enable the broader exploitation of non-food biomass resources.

What the United States does at home will have implications abroad. Successful development of an unconventional-fuel industry based on coal, biomass, and oil shale should promote similar developments in other nations that also have appreciable amounts of these resources. Considering both domestic and international opportunities, it is not unreasonable to assume that unconventional-fuel development, if pursued, could reduce world demand for crude oil by 15 million to 20 million bpd from what it would otherwise be after 2030.

CHAPTER FIVE

Benefits of Coal-to-Liquids Development

This chapter examines the strategic significance to the United States of developing a CTL industry. The research we report here builds on a previous RAND study (Bartis et al., 2005) that examined the related question of how oil shale development might contribute to achieving U.S. goals at home and abroad. For CTL development, this is an important issue because Congress is currently considering a number of legislative proposals to provide subsidies to promote the construction and operation of commercial CTL plants. Our analysis provides a context in which to evaluate the potential national benefits of such subsidies.

For the purpose of understanding the strategic significance of CTL development, we assume a future in which a mature CTL industry is operating without government subsidies in the United States and in other nations with large coal resources. Further, we assume that CTL production in the United States is displacing three million bpd of conventional oil production, which is our estimate for the maximum level achievable in the 2030 time frame, assuming successful commercial operations of a few early CTL and CBTL plants and timely demonstrations in the United States of capture and large-scale sequestration of carbon dioxide.

We benchmark our calculations of economic benefits using the EIA 2007 reference and high-oil-price cases for 2030 (EIA, 2007b). This implies a long-term petroleum price in the $60 to $100 per barrel range for a low-sulfur light crude oil.[1] Our analysis in effect looks forward at least 35 years (five years of CTL plant construction followed by 30 years of plant operation). The prices we use for the illustrative benefit calculations can be considered as long-term real prices averaged over this period.

World oil prices prevailing in early 2008 were well above this range. However, for two reasons, we have not extended upward the range of long-term oil prices we consider. The first is that, if long-term oil prices were to remain consistently above $100 per barrel, our analysis of technologies for CTL production would imply no need for government policy to help foster a successful commercial CTL industry. The second is

---

[1] The EIA projections for 2030 are $59.12 per barrel for its reference case and $100.14 for its high-oil-price case. These prices are in 2005 dollars. Adjusted to 2007 dollars, the projected reference and high-oil-case prices are about $63 and $106 per barrel, respectively.

that we also have doubts about such oil prices being sustained over the future period relevant to our analysis. These doubts stem from the following reasoning: If the high oil prices of 2008 persisted, there would be long-term, demand-moderating increases in the efficiency of oil use, particularly in transportation. Moreover, such prices would stimulate higher investment for new reserve development by producers not only of conventional oil but also of alternatives. Even if global production of conventional oil peaks during the period of our analysis, as some predict, many alternatives become viable in large quantities at long-term prices below $100 per barrel. It seems likely that the combination of new resources and greater energy efficiency will be sufficient to meet current and a growing future demand. Appendix C considers these points in more detail.

Economic Profits

A competitive CTL industry producing three million barrels of liquid fuels per day in the United States would displace about one billion barrels per year of a combination of imported crude oil and imported petroleum products. The value of this production would depend on world oil prices in 2030. Considering benchmark prices for low-sulfur light crude oil (similar to West Texas Intermediate) that range from $60 to $100 per barrel (in 2005 dollars; see EIA, 2007b), the annual value of the production from a three million bpd industry could range from $70 billion to $120 billion in 2007 dollars. In the absence of CTL production, most of this money would be spent on importing crude oil or finished petroleum products, as is currently the case.

If production costs, including a reasonable rate of return on capital investments, are below the prevailing market price for oil, a domestic CTL industry will generate economic profits. For example, if CTL production in 2030 costs an average of $40 per barrel[2] and world crude prices are between EIA 2007 reference-case and high-oil-price levels, the economic profits of a three million bpd domestic CTL industry would range from $20 billion to $70 billion per year (in 2007 dollars). Through income taxes, up to 35 percent of these profits ($7 billion to $25 billion) would go to the federal government and thereby broadly benefit the public.[3] In addition, through property, coal-severance, and income taxes, state and local governments would collect smaller amounts.

---

[2] By the time domestic CTL production has reached three million bpd, cumulative global CTL production will likely be well beyond ten billion barrels. The learning gained from this level of production should lead to significant reductions in CTL production costs below the $55- to $65-per-barrel range predicted for early commercial plants—hence our estimate of $40 per barrel as a potential cost for CTL derived from matured technology.

[3] Much of the coal located in the West is on federal lands. Federal, state, and local governments would also derive income from lease bonuses and royalty payments associated with mining on these lands. However, the total amount of the payments is likely to be much less than the amount obtained from federal income taxes. For

Reductions in the World Price of Oil

RAND researchers' earlier analysis of oil shale development (Bartis et al., 2005) found that a domestic oil shale industry would likely cause world oil prices to be lower than they would otherwise be. This would result in appreciable economic benefits to oil consumers in the United States and elsewhere. Additionally, we found that the primary national security benefits associated with domestic oil shale production would derive from lower world oil prices. While these findings were based on our analysis of oil shale development, they are more broadly applicable.

Based on fundamental economic considerations, we expect that lower world oil prices will likely be the result of any increase in liquid-fuel production, be it from conventional-petroleum extraction or from unconventional sources, such as coal, biomass, oil shale, tar sands, or stranded natural gas. Moreover, this effect is independent of whether the additional production is within the United States or in some other country. We also know that a reduction in world demand for oil through conservation or more efficient energy systems should likewise cause world oil prices to be lower than they would be in the absence of that reduction in demand.

These considerations suggest the possibility that production of unconventional fuels, combined with energy-conservation measures and gains in energy efficiency, might result in a substantially lower demand for conventional oil. Such a decrease could lead to a large reduction in world oil prices, increased economic benefits to oil consumers, and reduced revenues to OPEC. To estimate the size of these price reductions, benefits, and income flows, we developed a simple, transparent, empirically based parametric model of the world market for petroleum. This model is fully described in Appendix C. Because it will take at least 20 years to build an unconventional-fuel industry producing strategically significant levels of liquid fuels, our interest is on how unconventional-fuel production in the 2030 time frame would change world oil prices from what they would otherwise be.

At the heart of our world oil market model are assumptions regarding the potential behavior of the OPEC cartel. To bracket OPEC's behavior in response to changing world demand for oil, we take two extreme positions. At one extreme, we assume that a group of OPEC members (the OPEC core) will react cohesively to a long-term increase in non-OPEC fuel production or to a decrease in world oil demand.[4] In this case, we assume that core members reduce production to optimize profits. At the other extreme, we assume that the OPEC core cannot or does not react at all, with the result that the core members as a group maintain the same production levels as they would

---

example, the federal royalty payment for surface-mined coal is 12.5 percent of market value, implying a royalty payment of less than $2.00 per barrel of CTL fuel production.

[4]   The results presented in this chapter are based on an OPEC core consisting of Saudi Arabia, Kuwait, Qatar, and the United Arab Emirates. We assume that, given world market prices, all other members of OPEC produce to their full capabilities, i.e., they are price takers.

have done in the absence of an increase in non-OPEC fuel production or a decrease in world oil demand.

The analysis shows that world petroleum prices fall much more if the OPEC core breaks rank in response to reduced demand for its exports than if the OPEC core remains cohesive. And the response if the OPEC core is simply a price taker and pure price competitor falls between the two. So we believe this range reasonably brackets the most likely responses of OPEC to reduced demand.

Estimated World Oil Price Reduction

To estimate how much the world oil price might respond to a reduction in demand for petroleum, we calibrated our model using both the reference case and high-oil-price-case projections for 2030 published in 2007 by EIA (2007b). In its reference case, EIA assumes that world oil prices in 2030 will be $59 (in 2005 dollars) per barrel and forecasts world demand for liquid fuels to be 118 million bpd.[5] In its high-oil-price case, EIA assumes that world oil prices will be about $100 per barrel and forecasts world petroleum demand to be 103 million bpd. For both EIA cases, our analysis suggests that world oil prices would drop by between 0.6 and 1.6 percent for each million barrels of unconventional-liquid-fuel production that would not otherwise be on the market. The lower end of this range would be realized if the core members of OPEC are able to act cohesively and if both supply and demand for liquid fuels are more responsive to prices; and vice versa for the upper end of this range.

For liquid-fuel additions of up to ten million bpd, our analysis suggests that this price decrease should be close to linearly proportional to the increase in liquid-fuel production outside of OPEC.[6] For example, the 2030 maximum estimated level of CTL production in the United States (Table 3.1 in Chapter Three) of three million bpd of CTL (or combination of CTL and CBTL) fuels should cause world oil prices to drop by between 2 and 5 percent below what prices would otherwise be in the absence of such additional production. If domestic CTL production were to be accompanied by an equal amount of CTL production in other countries, the world oil-price drop would double to between 4 and 10 percent.

The magnitude of the potential world oil-price decline depends on where crude oil prices are heading over the longer term. Figure 5.1 illustrates this for each one million bpd of additional CTL fuel production. For example, applying a three million bpd increase to EIA's 2007 reference case of about $60 per barrel results in a price decrease

---

[5] The $59 per barrel price is for a low-sulfur, light crude oil similar to West Texas Intermediate. The average price of all crude oil imports is significantly lower. For example, for its reference case, EIA assumes that average delivered prices of crude oil imports to the United States will be $51.63 per barrel in 2030.

[6] The results of the world oil market model are based on the assumption of a linear response between liquid-fuel quantities and prices. Economists sometimes prefer log-linear responses. Over the range of unconventional-fuel-supply additions that we examined (up to ten million bpd), our results do not significantly change if we assume log-linear supply and demand curves (see Appendix C).

Figure 5.1
Estimated World-Oil-Price Decrease for Each One Million bpd of Unconventional-Liquid-Fuel Production

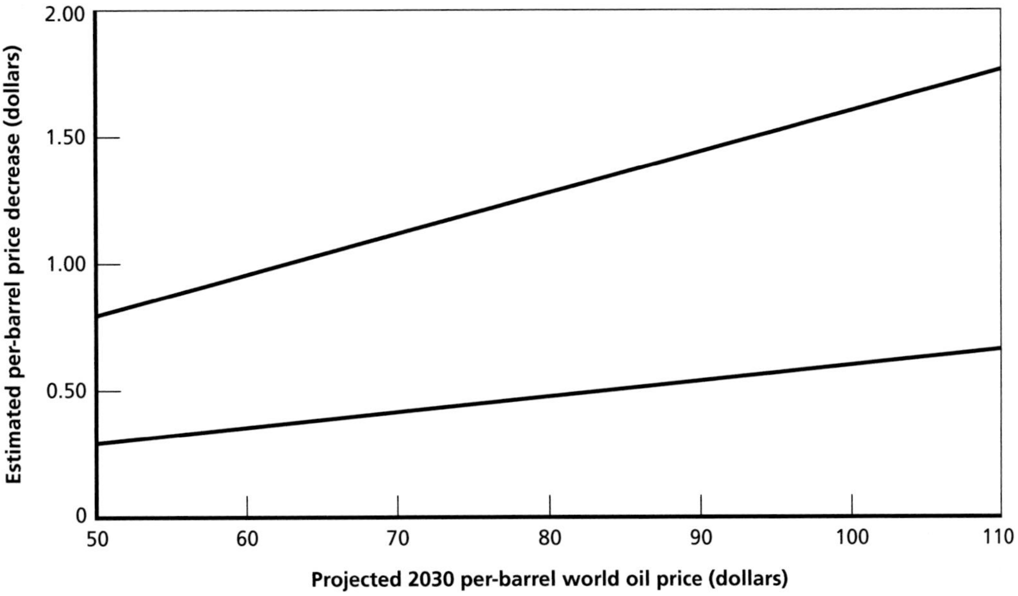

NOTE: The projected 2030 world oil price is that price that would be in effect if the unconventional-fuel production capacity were not built.
RAND *MG754-5.1*

of between $1.10 and $2.90 per barrel. Likewise, for EIA's 2008 reference case of about $70 per barrel, the price decrease would be between $1.30 and $3.40 per barrel, and, for EIA's 2007 high-oil-price case of about $100 per barrel, the price decrease would be between $1.80 and $4.80 per barrel.

Benefits and Costs to Oil Consumers and Producers in the United States

The anticipated reduction in world oil prices yields important net economic benefits. In particular, if aggressive investment in CTL development occurs and 2030 production reaches three million bpd, U.S. consumers would pay $7 billion to $59 billion less per year for liquid fuels, as shown in Table 5.1, which is based on calculations documented in Table C.2 in Appendix C (escalated to 2007 dollars).[7]

Some of this benefit to consumers would come at the expense of U.S. producers of petroleum and close petroleum substitutes, including unconventional fuels. The reduction in prices induced by an additional three million barrels of unconventional

---

[7] To be consistent with our model, the effects shown in Table 5.1 are based on the average price of imported crude oil delivered to U.S. refineries in 2030, which EIA estimates (in 2005 dollars) to be $51.63 for the reference case and $92.93 for the high-oil-price case.

Table 5.1
Calculated Changes in U.S. Consumer, Producer, and Net Surplus in 2030 Attributable to Unconventional-Fuel Production of Three Million Barrels per Day

Annual Surplus	EIA Reference Case (billions of 2007 dollars)	EIA High-Oil-Price Case (billions of 2007 dollars)
Consumer	7 to 32	12 to 59
Producer	−2 to −11	−6 to −27
Net	4 to 21	7 to 32

fuel would reduce the net income of such producers by about $2 billion to $27 billion per year (Table 5.1, second row).

Note that these losses are comparable in character, albeit smaller, to the potential gains in economic profits, discussed at the beginning of this chapter, from investments in CTL and other unconventional liquid fuels. However, they reflect different effects and affect different producers. The increased economic profits just discussed accrue directly to firms that invest in new unconventional-liquid-fuel production through standard market mechanisms. The loss in producer surplus affects all other U.S. producers of liquid fuels, because the new unconventional-fuel production will drive down world prices. Standard market mechanisms do not punish new producers for imposing such a loss on existing producers. Both effects are important to our total analysis. New production can yield increased profits for some firms and lower profits for others, effectively imposing a transfer of income between these producers.

The net gain to the U.S. economy from lowering oil prices is the difference, shown in the third row of Table 5.1, about $4 billion to $32 billion a year. The lower end of this range would hold if (1) 2030 world oil prices are near $60 per barrel, (2) the OPEC core can act cohesively and optimize its revenues, and (3) the supply of and demand for liquid fuels from outside of the OPEC core is responsive to prices. The higher end of this range would hold if (1) world oil prices are near $100 per barrel, (2) OPEC cannot act cohesively, and (3) the supply of and demand for liquid fuels from outside of the OPEC core are less responsive to prices. Given that the extreme circumstances we considered are unlikely to occur in all three dimensions, we suggest, for policymaking purposes, $6 billion to $25 billion per year as a more reasonable range for the net gain to the U.S. economy from the world oil-price decrease that would result from the production of three million bpd of CTL or, for that matter, any other unconventional fuel.

Just as the world oil-price decrease occurs no matter whether CTL production occurs within the United States or abroad, the benefits to U.S. consumers and to the U.S. economy occur no matter where CTL production takes place. In particular, the results shown in Table 5.1 pertain to a global production level of three million barrels of CTL fuels. While we estimate in Chapter Three that three million bpd is roughly the maximum level of CTL production that can be achieved by 2030 in the

United States, comparable or even higher production levels might be reached abroad by that time. As discussed in Appendix C, the consumer, producer, and net economic surpluses are close to linear functions of CTL production. A global CTL production level of six million bpd would roughly double the estimates of U.S. surplus gains shown in Table 5.1, while a production level of 1.5 million bpd (resulting from less aggressive CTL deployment) would roughly halve the estimates.

Benefits per Coal-to-Liquids Barrel

If three million bpd of unconventional-liquid-fuel production results in the net benefits shown in Table 5.1, each single barrel produced should have an appreciable benefit. The marginal impacts are nonlinear, but only slightly, depending weakly on how much new unconventional-liquid-fuel production would occur before that additional barrel would be added. Table 5.2 reports our results (from Table C.2 in Appendix C, adjusted to 2007 dollars), assuming that the single extra barrel is on top of three million bpd of production. At the margin defined by production of an additional three million bpd of unconventional fuel, an additional barrel of unconventional fuel would increase the consumer surplus by $6 to $55, with the range dependent on the future oil prices, OPEC cohesion, and the price response of crude oil supply and demand. The caveats regarding the extreme circumstances just discussed suggest that a more reasonable range for the consumer surplus is $10 to $35 per barrel of produced unconventional fuel. From the perspective of liquid-fuel consumers in the United States, they reap benefits from government policies that provide a $10 to $35 per barrel advantage to unconventional fuels as compared to conventional-petroleum products.

As before, a portion of this benefit to consumers in the United States would come at the expense of liquid-fuel producers in the United States. After subtracting the loss to domestic producers, a significant marginal net surplus remains, as shown in the bottom row of Table 5.2. For policy decisionmaking, a more reasonable range for the net marginal surplus is $6 to $24 per barrel.

These values can be viewed as a measure of the classic oil-import externality. The United States should be willing to spend $6 to $24 per barrel more than market prices for substitutes that reduce the demand for oil production and thereby put downward

Table 5.2
Marginal Changes in U.S. Consumer, Producer, and Net Surplus Attributable to Unconventional-Fuel Production

Marginal Surplus per Barrel of New Production	EIA Reference Case (2007 dollars per barrel)	EIA High-Oil-Price Case (2007 dollars per barrel)
Consumer	6 to 30	11 to 55
Producer	−2 to −10	−5 to −24
Net	4 to 21	6 to 30

pressure on world prices to counter efforts by OPEC members to hold prices above competitive market levels by restraining production.[8] This premium above market prices is not particularly sensitive to the number of barrels actually produced; therefore, it remains valid over a wide range of levels of new production.

National Security Benefits

The anticipated reduction in world oil prices associated with CTL development, or more generally with unconventional-fuel development, may also yield national security benefits in addition to the economic benefits just discussed. In 2007, OPEC revenues from oil exports were about $700 billion per year.[9] Projections of future petroleum supply and demand published by the U.S. Department of Energy (EIA, 2007b, 2008c) indicate that, unless measures are taken to reduce the price of and demand for OPEC petroleum, such revenues will grow considerably. These high revenues raise several national security concerns because some oil-exporting countries are governed by regimes that are not supportive of U.S. foreign policy objectives. Income from petroleum exports has been used by unfriendly nations, such as Iraq under Saddam Hussein and Iran, to support weapon purchases or to develop their own industrial base for munition manufacture.

Also, the higher prices rise, the greater the chances that oil-importing countries will pursue special relationships with oil exporters and defer joining the United States in multilateral diplomatic efforts. For example, in 2004, Japanese firms signed an agreement with the National Iranian Oil Company to undertake the development of Iran's Azadegan oil fields, which would be the largest oil-development project ever attempted by a Japanese firm. This project was pursued in the name of energy security—and, with the concurrence of Japan's foreign ministry, in the name of better bilateral relations with Iran. Japan's negotiations with the Iranian government came into conflict with the U.S. government's efforts to compel Iran to abide by its International Atomic Energy Agency obligations. Japan was not dissuaded from concluding the Azadegan agreement. Implementation of the Azadegan agreement is now on hold, but only as a result of greatly increased international concerns regarding Iran's nuclear program and intentions.

Most of the major oil-exporting nations have serious governance problems, the notable exceptions being Canada and Norway. It is difficult, however, to find broadly based relationships between these governance problems and oil revenues. On the one

---

[8]   This range is compatible with those for other recent estimates of this value. For an especially useful review of the literature, see Leiby (2007).

[9]   This estimate is based on 2007 OPEC liquid production of 35.4 million bpd (EIA, 2008d, Tables 11.5, 11.6), our estimate of 2007 OPEC liquids consumption of about eight million bpd, and average 2007 U.S. refinery acquisition costs for imported oil of $67.02 per barrel (EIA, 2008d, Table 5.21).

hand, research suggests that high oil revenues allow some regimes to resist democratization or forgo needed internal social and economic reforms (Klare, 2002; Ross, 2008), as appears to be the present case in Russia and Venezuela. Moreover, the perception that the United States may be supporting an authoritarian regime can lead to anti-U.S. sentiments and possibly acts of terrorism against the United States.

On the other hand, arguments can be put forth that low oil prices can destabilize governments or cause social and economic disruptions that harm the national security of the United States. For example, one might argue that the very low oil prices of the 1990s hampered Russia's attempts at social and political reform, or that low oil prices in the late 1980s and 1990s contributed to the social unrest in Saudi Arabia that spawned al Qaeda. Moreover, major investments in education and progress in social reform that are currently occurring within certain oil-exporting nations—such as Qatar, the United Arab Emirates, and Oman—might have been catalyzed by current high oil prices.

Our calculations show that domestic CTL production of three million bpd would reduce annual OPEC oil revenues by between 3 and 8 percent by 2030. This small a change in revenue would unlikely change the political dynamics in oil-producing nations that are unfriendly to the United States. However, if domestic CTL production can be combined with other economical unconventional-fuel production, appreciable revenue reductions could result. For example, global unconventional-fuel production of ten million bpd will reduce annual OPEC oil-export revenues by between 11 and 26 percent (see Appendix C, Table C.3), the smaller reduction being associated with lower world oil prices and price elasticities that yield smaller world oil-price reductions. In 2007 dollars, these reductions amount to between $98 billion and $300 billion per year. As before, this range includes combinations of extremes in world oil prices, elasticities, and OPEC behaviors that are unlikely to simultaneously occur, so that a somewhat narrower range may be more appropriate for policy deliberations.

We close this section with the important caveat that it is only to the extent that lower oil prices encourage favorable foreign policies or democratization and social and economic reforms in oil-exporting countries that benefits accrue to the national security of the United States. Moreover, favorable impacts, from the perspective of the United States, on certain oil-exporting nations could be accompanied by adverse impacts on others.

Improved Petroleum Supply Chain

Due to the broad regional dispersion of the U.S. coal resource base and the fact that CTL plants are able to produce certain finished fuel products that are ready for retail distribution, developing a CTL industry should increase the resiliency of the overall petroleum supply chain. In particular, a CTL industry will likely reduce the fraction of finished fuels that are produced in refineries located in Texas, Louisiana, and Mississippi that are vulnerable to operational disruptions from hurricanes and other adverse weather events. Also, a domestic CTL industry should reduce U.S. dependence on imports of finished petroleum products. In particular, imports of highly volatile products (e.g., automobile gasoline) are projected to continue over the next 20 years (EIA, 2008c, Table A11). Reducing such imports would result in a significant decrease in the amount of hazardous material entering U.S. ports and waterways.

The major finished product from an MTG CTL plant is gasoline that can be sent to local distribution terminals. For FT CTL plants, the major finished products that can be directly distributed are diesel fuels suitable for highway and off-road uses, jet fuels suitable for commercial and military applications, and fuel oils suitable for home heating and certain industrial applications. For the first few FT CTL plants, we anticipate that the naphtha cut, which will likely be between one-fourth and one-third of total CTL production, will be transported to existing petrochemical plants, where it can be used to produce ethylene and related chemical feedstocks. If a large CTL industry develops, it is likely that centralized facilities for naphtha upgrading or new refineries specifically designed to accept both crude oil and CTL naphtha will be constructed. In either case, the result will be greater geographic diversity in the sources of automotive and aircraft fuels.

Oil-Supply Disruption Benefits

No matter what the cause, a sudden and large disruption in world oil supplies could have significant economic consequences to the United States and other oil-importing nations. In general, a sudden loss in oil supplies to the world market would lead to a rapid increase in world prices, otherwise known as a *price shock*. The magnitude and duration of an oil price shock will depend on the underlying causes and on the effectiveness of national and international efforts to mitigate supply disruptions.

An important lesson learned from the experiences of the 1970s in responding to oil price shocks is the importance of market-induced conservation as opposed to a government-controlled energy-allocation system, which led to the closing of factories and the notorious gasoline lines of the 1970s. But by increasing the costs of energy consumption, a market-based response policy places a large burden on energy consumers, especially those in moderate-income households, who could suffer a large loss in purchasing power.

The less dependent the economy is on oil, the less the adverse impacts of an oil price shock on industry and final consumers. For this reason, energy policies that, over the long term, encourage decreased dependency on petroleum via efficiency and conservation will reduce the adverse impacts on any oil price shock. But energy policies that promote continued or increased dependence on liquid fuels can be ineffective or counterproductive in mitigating the adverse effects of oil price shocks. For example and as discussed earlier, the development of a competitive domestic CTL industry will cause world oil prices to be below what those prices would be in the absence of a CTL industry. Reduced prices promote increased demand, with the net result being an economy that is more dependent on liquid fuels and thereby more vulnerable to a sudden supply disruption. After all, if world oil prices increase, domestic liquid-fuel producers, including the operators of CTL plants, will raise their prices to match the new competitive range, as they should if market-induced conservation is to occur.

Nonetheless, having a larger domestic liquid-fuel production industry provides the government with greater policy options than it would otherwise have during an oil-supply disruption. In particular, an oil price shock results in large (potentially, very large) wealth transfers from oil consumers to oil producers. Having a larger domestic oil production industry means that more of these excess profits would be generated on U.S. soil, where they could, for example, be heavily taxed and redistributed to oil consumers. However, it is questionable whether effective legislation to address the wealth-transfer issue can be enacted to achieve an equitable distribution, especially in a crisis atmosphere, without impairing market-induced deployment of additional domestic or international oil production capacity.

Employment Benefits

The development of a domestic CTL industry will cause new jobs to be created, especially in those states in which CTL plants would be built. These are likely to be the states holding the largest fraction of U.S. coal reserves listed in Table 2.2 in Chapter Two. Employment opportunities are currently limited in several of these states, and many members of their governments view CTL development as supportive of state and regional economic development.[10]

The broader effect of CTL development on national employment is less clear. If macroeconomic factors are such that the nation is near full employment, which has been the case for most of the last two decades, the labor demands associated with CTL development will be met either by declining employment in other sectors of the economy or by additional immigrant labor.

---

[10] During the course of our research, this view of the relationships among CTL development, employment, and economic development was expressed during meetings with Governor Joe Manchin III of West Virginia (June 2007), Governor David D. Freudenthal of Wyoming (May 2007), and members of the Kentucky legislature (May 2007).

Confounding or Inconclusive Arguments

In the previous study on oil shale, we assessed and found weak the following three arguments that proponents of increased domestic production of energy supplies often raise to justify federal support or involvement. We find that these arguments are also weak in the context of CTL development.

1. *Unconventional-fuel development will reduce the current account deficit.* There is a favorable impact, but it is indirect and could be small. In a full-employment economy, which has characterized the U.S. economy for most of the past two decades, the development of a domestic CTL industry would result mainly in the shift of labor from other sectors to the new industry. Although imports of crude oil would fall, the accompanying decline in output from other sectors could cause net imports of all goods and services to increase. Assuming that net U.S. borrowing from abroad remained constant, overall levels of exports and imports would remain unchanged (Krugman and Obstfeld, 2002). However, because domestic CTL production should lead to a decline in the world price of oil and therefore an improvement in the terms of trade for the United States, the trade balance would improve over the medium term. With lower aggregate borrowing costs, the trade balance could also improve in the long run as the stock of U.S. liabilities abroad would not have increased as much, and aggregate U.S. spending on imported oil would be less.

2. *Eliminating or decreasing U.S. dependence—but not the dependence of the rest of the world—on Persian Gulf sources of petroleum will increase the national security of the United States and allow the United States to significantly or even drastically reduce its engagement in the Middle East.* This argument does not recognize that the United States is a participant in a global marketplace for energy supplies. Whether or not the United States imports oil from the Persian Gulf, a shortfall in exports from that region would have the same net effect on the overall supply of oil to the United States and the prices paid by U.S. consumers. For example, the United States does not import any oil from Iran, yet a sudden reduction in Iranian exports would cause a global increase in oil prices, including prices paid in the United States for both domestic and imported oil. This argument also ignores the fact that the United States is deeply engaged in the Middle East for a variety of reasons and cannot isolate itself from continuing threats based on religious, political, and economic strife. Less than 15 percent of the petroleum exports of the Persian Gulf nations land in the United States. This relatively small U.S. share of Persian Gulf oil exports is not proportional to the current level of U.S. engagement, including the U.S. relationship with Israel and the predominant role of the United States in both Iraq and Afghanistan. It is true that the United States has assumed the major role in protecting the petroleum

supply chain that begins in the Persian Gulf. But this effort does not reflect the relatively small share of Persian Gulf exports headed to the United States.

3. *Greater domestic production of liquid fuels will increase the reliability of fuel supplies for the U.S. military.* Given that the U.S. military is responsible for only about 1.5 percent of total U.S. consumption of petroleum,[11] it is inconceivable that, in the foreseeable future, the armed forces will not be able to purchase on the open market all the fuel required to meet their needs.[12] The question is how much they will have to pay, not whether they can find adequate supplies.

The Economic Value of a Domestic Coal-to-Liquids Industry

The total economic value of a domestic CTL industry consists of the value of each of the benefits described in the preceding section minus any environmental or other social costs. To develop a rough estimate of this value, we consider only economic profits and the net economic surplus and examine how these two measures accumulate as a four million bpd domestic CTL industry develops at a fairly rapid pace.[13] Assuming that an average crude oil price of $100 per barrel will hold over the calculation period and a high value ($24 per barrel, Table 5.2) for the net economic surplus, we calculate the net present value of national benefits to be roughly $500 billion (2007 dollars).[14] If we assume that a lower average oil price ($60 per barrel) will hold and consider a lower value for the net economic surplus ($6 per barrel), the net present value of national benefits decreases to roughly $140 billion.

The preceding estimates are based on the assumption that a decision to build the first round of commercial CTL plants would be made in 2008. Delaying the onset of commercial CTL development would cause these estimates to drop significantly. For example, a five-year delay in obtaining early commercial experience would drop these values by about 30 percent, and a ten-year delay would erode projected benefits by about 55 percent.

---

[11] Total U.S. Department of Defense (DoD) petroleum use in 2007 was 0.67 quads (EIA, 2008c, Table 1.13), which is about 330,000 bpd. Total U.S. consumption has been about 21 million bpd.

[12] A 2006 JASON study commissioned by the director for defense research and engineering at DoD (Dimotakis, Grober, and Lewis, 2006) also reached this conclusion.

[13] We assume that CTL production grows at the rate shown in Table 3.1 in Chapter Three (for the case of accelerated carbon capture and sequestration demonstration) to four million bpd by 2033 and thereafter remains constant. Present values are calculated using the Office of Management and Budget (OMB)–prescribed 7-percent real discount rate through 2050.

[14] The net present value calculations cover CTL production through 2060 and are based on CTL production costs (including return on capital investment) of $60 per barrel crude oil equivalent (2007 dollars) for early plants, progressing to $40 per barrel by 2030 through experience-based learning.

The national security, petroleum supply chain, and regional benefits associated with a domestic CTL industry are difficult to quantify and, for that reason, are not included in this estimate. So long as CTL industrial development includes carbon capture and sequestration, environmental and social costs should not significantly detract from these values.

Our estimates of the economic value of a domestic CTL industry are very sensitive to world oil prices, especially those holding beyond 2025. Higher world oil prices imply a higher net social surplus and higher economic profits and would yield a higher overall economic value to the nation, as compared to the estimates just given. Another assumption driving our results is the extent of CTL cost reduction that will be achieved through learning after initial operating experience is achieved. For example, if ultimate CTL production costs are $50 as opposed to $40 per barrel crude oil equivalent, the net present value of estimated national benefits for the high-oil-price case (namely, about $500 billion) would drop by about $40 billion. We also emphasize that our estimates of national benefits are based on a domestic CTL industry that is developing at or close to what we consider to be the maximum feasible rate, as discussed at the conclusion of Chapter Four. A slower rate on CTL industrial growth would cause our estimates to decrease.

CHAPTER SIX

Critical Policy Issues for Coal-to-Liquids Development

Based on our research and interviews with major energy producers, market and technical uncertainties affecting the overall commercial viability of CTL production appear to be the primary reason for delayed investment in CTL plants. Absent some bounding of these uncertainties, development of a strategically significant CTL industry in the United States is unlikely to proceed. Consequently, over the next five to ten years, the prospects for CTL development in the United States depend largely on the extent to which the federal government will provide incentives or disincentives for private investment in early commercial CTL production facilities.

Another class of uncertainties affecting CTL development is how Congress and the states will view such investment decisions in light of concerns about greenhouse-gas emissions and other anticipated environmental impacts that would result from a large-scale CTL industry. The trajectory of CTL development will be affected by how existing environmental regulations are implemented and enforced, whether changes in those regulations are made, and whether new regulatory or market-based mechanisms to reduce emissions and other land, water, and biological impacts are authorized. These issues are explored in this section.

Environmental Impacts of Coal-to-Liquids Production

Greenhouse-Gas Emissions

As a consequence of ongoing assessments of the potential adverse impacts of global climate change,[1] Congress is considering several alternative measures to reduce emissions of greenhouse gases. States are also engaged in this debate, with California already taking the boldest move through its passage of a law mandating emission reductions (State of California, 2006). Most proposals related to energy development at the federal, state, and local levels are now being viewed through the prism of concerns about

---

[1] Most notably, the Fourth Assessment Report of the Intergovernmental Panel on Climate Change (2008), especially the assessment of climate-change impacts and vulnerability (IPCC, 2007a).

global climate change and emissions of carbon dioxide from burning fossil fuels (Barringer and Sorkin, 2007).

In the United States, annual greenhouse-gas emissions from human activities, measured as carbon dioxide equivalents, are about seven billion tonnes per year (EIA, 2008d, Table 12.1) with carbon dioxide dominating at about six billion tonnes per year.[2] An international consensus exists that the only effective way to begin reducing greenhouse-gas emissions and slow global climate change would be to associate a cost with emitting carbon dioxide and other greenhouse gases through some type of policy instrument, whether a market-based mechanism (such as a carbon tax or a cap-and-trade emission program) or direct regulation. Without such a price tag, businesses and consumers will not make the investments in technology and changes in energy use that are required to dramatically reduce those emissions.

In Chapter Three, we showed that carbon dioxide capture from a CTL plant should be considerably less expensive than capture from a new coal-fired power plant. Consequently, any market-based greenhouse-gas control program sufficient to promote carbon capture and sequestration at a new coal-fired power plant should be more than sufficient to promote capture and sequestration at a CTL plant.

In the absence of an *effective* national program to reduce greenhouse-gas emissions, it is unclear whether the federal government would support the development of a CTL industry capable of producing millions of barrels of liquid fuels per day. For example, if no legal regime for carbon management existed and CTL development were to proceed, a CTL industry capable of producing three million barrels of liquid fuel per day and operating without carbon-management measures would annually emit between 700 million and 850 million tons of carbon dioxide.[3]

Large-scale carbon sequestration has not yet been demonstrated in the United States. U.S. Department of Energy plans for demonstrating large-scale carbon-sequestration center on the Regional Carbon Sequestration Partnerships. Managed by the National Energy Technology Laboratory, this partnership program may result in the start-up of eight or more moderate- to large-scale demonstrations over the next five years (see boxed insert). These demonstrations are intended to provide information required for selecting sequestration sites (including associated federal, state, and local regulation and permitting activities), assessing permanence of storage, designing injection and monitoring systems, determining overall economics, and reducing public uncertainties regarding risks. Once the viability of carbon sequestration is established, CTL production and use should be possible with net greenhouse-gas emissions that are commensurate with or slightly below those of conventional-petroleum products.

---

[2] The other significant greenhouse gases are methane, nitrous oxide, hydrofluorocarbons, perfluorocarbons, and sulfur hexafluoride.

[3] The lower part of this range assumes that significant performance improvements will result as CTL production experience is achieved, as discussed in Chapter Three.

The Regional Carbon Sequestration Partnerships

The Regional Carbon Sequestration Partnerships consist of seven government and industry efforts to determine the most suitable technologies, regulations, and infrastructure needs for carbon dioxide sequestration in different areas of the country. The partnerships include participation from state agencies, universities, and private companies covering 41 states and four Canadian provinces. On the federal side, the partnership program is supported and coordinated by the National Energy Technology Laboratory.

Starting in 2003, the initial phase of work by the partnerships centered on developing predictive models to access available geologic storage capacity within each region. The second work phase centered on in-field characterization research and small-scale testing of carbon dioxide injections. The objectives of the phase II work are further development and validation of mathematical models that describe the interaction of carbon dioxide with key geologic formations in each region and an estimate of regional carbon dioxide storage capacity.

The partnership program is now transitioning to its third phase, which consists of moderate- to large-volume injections. In late 2007, the U.S. Department of Energy announced that four partnerships—Plains CO2 Reduction Partnership, Southwest Regional Partnership on Carbon Sequestration, Southeast Regional Carbon Sequestration Partnership, and Midwest Regional Carbon Sequestration Partnership—would receive $264 million in federal funding to demonstrate carbon dioxide sequestration in saline formations at five sites in the United States, at injection rates equivalent to 100,000 to one million tons per year, and one site in British Columbia, at an injection rate of 1.8 million tons per year. Carbon dioxide injections would continue over a three- to six-year period, depending on the project. The remaining three regional partnerships (Big Sky Regional Carbon Sequestration Partnership, Midwest Geological Sequestration Consortium, and West Coast Regional Carbon Sequestration Partnership) are anticipated to transition to phase III during 2008.

Other approaches, such as CBTL, to achieve or go slightly beyond parity with conventional petroleum were described in Chapter Three. By using a combination of carbon sequestration and cofeeding with coal and biomass, it is possible to reduce total fuel-cycle carbon dioxide emissions to levels that are well below those of conventional petroleum (NETL, 2007e) or to even reduce the amount of carbon dioxide in the atmosphere, as was shown in Figure 3.4 in Chapter Three.

Once large-scale carbon sequestration is proven possible, there is still the issue of whether the geologic formations in the United States are sufficient to allow cost-effective and safe annual sequestration of three or more billion tons of carbon dioxide, which would be the case if a high level of carbon capture were to be required at CTL plants and fossil-fuel power plants. It may be that the development of a large domestic CTL industry is not compatible with growing dependence on coal for electric-power generation. While the coal resource base appears adequate to provide fuel for both power generation and significant CTL production, the availability of suitable sites for carbon sequestration may be the limiting factor over the longer term.[4]

---

[4] The IPPC (2005) carbon dioxide storage-capacity estimate is 675 billion to 900 billion tonnes in depleted oil and gas reservoirs and at least one trillion tons (and possibly an order of magnitude more) in deep saline formations. These are global estimates. For comparison, current global carbon dioxide emissions are 24 billion tonnes per year.

Although the technical development of geologic carbon sequestration would substantially reduce, if not eliminate, greenhouse-gas emissions, the process of transporting and sequestering carbon dioxide could, if unmitigated, involve adverse environmental impacts. A recent review of the potential environmental effects associated with carbon sequestration identified risk factors that could cause moderate to significant adverse impacts to groundwater resources and human health and safety (NETL, 2007d).[5] These impacts appear to be associated with improper site selection or site preparation. More information on environmental impacts, as well as on prevention and mitigation measures, should be forthcoming as further technical progress is achieved, especially during the planning and design of moderate- and large-scale demonstrations, such as those occurring under the Regional Carbon Sequestration Partnerships. Such planning will involve extensive engineering and safety and environmental studies needed to obtain required construction and operating permits, address potential legal liabilities, and comply with the processes and environmental assessments required by the National Environmental Policy Act.[6]

Air Quality

Current federal and state environmental laws regulate the release of pollutants from coal mining and coal-fired generation plants. Covered pollutants include those currently designated by the U.S. Environmental Protection Agency as criteria pollutants (namely, sulfur oxides, nitrogen oxides, particulates, ozone precursors, and carbon monoxide)[7] as well as noncriteria pollutants that are currently on the list of air toxins covered by the federal Clean Air Act.[8] The law permits releases, but only below regulated levels. Presumably, CTL plant operations would be subject, at a minimum, to the same regulatory controls. Methane released during coal mining, not currently regulated, is not only a greenhouse gas but also a known precursor of ozone formation. Of the air toxins, mercury emissions are of special concern because coal contains trace amounts of mercury and coal combustion is currently a major source of mercury emissions.

The front-end and electric-power-production portions of an FT or MTG CTL plant are very similar to proposed "clean-coal power plants" based on IGCC power generation. The principal difference is that the CTL plants would incorporate more extensive removal of both mercury and sulfur to protect the catalyst used in the synthesis reactor. Analyses of IGCC power-generation (NETL, 2002; Nexant and Cadmus

---

[5]   A significant adverse environmental impact is defined (NETL, 2007d) as an impact that would be long term and widespread and that, despite mitigation measures, would result in violation of environmental statutes and regulations.

[6]   Public Law 91-190, National Environmental Policy Act, January 1, 1970; 42 U.S. Code 4321–4344.

[7]   40 Code of Federal Regulations 50, National Primary and Secondary Ambient Air Quality Standards.

[8]   42 U.S. Code 7412, Hazardous Air Pollutants.

Group, 2006) and FT CTL plants (NETL, 2007b; SSEB, 2006, Appendix D) indicate that sulfur oxide emissions can be controlled at a level of at least 99 percent, mercury emissions at a level of at least 95 percent, and particulate emissions at a level of less than 0.1 pound per ton of coal input. With these high levels of control, sulfur oxide, mercury, and particulate emissions from a CTL plant are unlikely to prevent the successful permitting of a CTL plant.[9]

The only significant sources of nitrogen oxide emissions are the gas turbines used to produce power used within the CTL plant and sold to the grid. The amount of fuel consumed by the gas turbines can vary significantly based on how the CTL plant is designed. A reasonable range for a CTL plant is for 70 to 140 megawatts (MW) of gas-turbine capacity to be in operation for each 10,000 bpd of liquid-production capacity. Nitrogen oxide emissions from these units should be comparable to the state of the art for turbines designed for combined-cycle power plants. Much lower levels of nitrogen oxide emissions can be achieved by using a selective catalytic reduction system.

A CTL plant is a complex facility and affords a number of unit operations that are potential sources of emissions of hazardous or noxious (e.g., odorous) air pollutants. Although some of these sources are unique to coal processing, most are associated with operations encountered in petrochemical facilities. Ensuring that such releases do not occur or are within acceptable limits is an inherent part of the environmental permitting process and front-end engineering design and detailed design of a CTL plant.

During the next few decades, air-quality concerns are not likely to seriously impede the development of a domestic CTL industry. The technical advances in pollution control, prevention, and monitoring that have been made in the past 30 years and still continue should allow the construction and operation of CTL plants without threat to human health or the environment. While there are surely potential plant sites at which local air-pollution issues may preclude construction, it is reasonable to expect that the geographic diversity of the U.S. coal resource base will be sufficient to provide adequate sites to support an industry that produces strategically significant amounts of liquid fuels.

Nevertheless, over the longer term—beyond 2030—the need to maintain air quality could lead to constraints on CTL production, especially in Wyoming and Montana. The coal resource base in these states is adequate to support ten to 20 or more large CTL plants. However, such extensive production may not be compatible with prevention of significant deterioration of air quality, a key requirement of the federal Clean Air Act, especially in nearby national park and wilderness areas, such as Badlands National Park.[10]

---

[9] For example, at 99-percent sulfur removal, a three million bpd CTL industry would release less than 300,000 tons (sulfur dioxide equivalent) per year of sulfur oxides into the atmosphere, as compared to current emissions from fuel combustion of more than 12 million tons per year (EPA, 2007a).

[10] 42 U.S. Code 7491, Visibility Protection for Federal Class I Areas.

Land Use, Ecological Impacts, and Water Quality

Plant-Site Impacts. Consistent with regulations and modern engineering practice, CTL plants will be built with zero discharge of water. Plant-site threats to water quality would be associated with the management of the solid wastes generated by the plant.

The most hazardous solid wastes will likely be generated from the evaporation of cooling and process water. These wastes will contain toxic metals and will need to be handled as hazardous wastes, per applicable regulations. In terms of bulk, the major waste product generated at the plant site will be coal ash. Coal ash is formed by the inorganic constituents of coal, which generally represent between 5 and 15 percent of the weight of coal. Assuming 10 percent, a 30,000 bpd CTL plant would generate about 0.5 million tons per year of ash. Coal gasification generally occurs at sufficiently high temperatures to cause most of this ash to be produced as a slag, an amorphous, glassy substance. Compared to coal ash, gasification slag is much less susceptible to leaching when placed in contact with water; therefore, its disposal presents lower hazards to groundwater resources. Gasification slag has been used as an additive for concrete production and shows promise for use in road and construction materials.

The footprint of a moderate- to large-sized CTL plant will be fairly large—roughly 0.5 square miles—comparable to that of a major petroleum refinery. In general, coal-mining operations occur in rural areas, some of which have scenic, recreational, or other uses. As a result, there will be locations where construction and operation of a CTL plant will be problematic. Again, given the geographic diversity of the U.S. coal resource base, lack of suitable plant sites is unlikely to impair large-scale CTL development. Moreover, in light of the size of the investment required to construct a CTL plant, the dedication of extensive resources for mitigating environmental and other ecological impacts of diverting land use to CTL production should not significantly affect overall plant economics.

Coal-Mining Impacts. For CTL development, the primary environmental issue at the regional and local levels will likely be the incremental impacts of increased coal mining on land use, the local ecology, and water quality. Adverse safety and environmental impacts of coal mining—even with regulation—are well documented and include mine drainage, mine fires, waste piles, ground movements (subsidence), and hydrological impacts (NRC, 2007).

For example, a prominent environmental issue associated with coal mining centers on mountaintop removal, such as that conducted in the steep terrain of the central Appalachian coalfields. This method can involve removing as much as 500 feet from a mountain summit in order to get to the coal seams underneath. The dirt taken from the mountaintop is placed in adjacent valleys, where it can cause the diversion of existing streams. According to the U.S. Environmental Protection Agency, blasting operations associated with mountaintop removal have cracked the foundations and walls of nearby houses, caused wells to dry up or become contaminated, damaged aquifers, destroyed native plants, and caused loss of habitat for wildlife (EPA, 2007a; EPA, U.S. Army Corps of Engineers, et al., 2005).

Another difficult issue is the management of coal-slurry impoundments, which contain wastes primarily from the processing and cleaning of coal at mines. Accidental breaches have caused loss of life and economic and ecological damage, thus raising questions about their physical integrity in the near and long terms.[11]

Resolving such issues as these is a matter of national as well as regional and local concern. It will require research and ultimately may require changes to existing regulations. If large-scale development of a CTL industry is accompanied by a significant net increase in coal production or a significant change in extraction technologies, a review of the legislation and regulations governing mine safety, environmental protection, and reclamation may be appropriate. Such a review would assess the potential environmental and safety impacts of increased mining activity and evaluate options for reducing such impacts. More immediately, there is a clear need for research directed at mitigating the known and anticipated environmental impacts and reducing the work hazards associated with coal mines, as recommended by the National Research Council (2007).

Water Requirements

In FT or MTG CTL plants, water is consumed during coal gasification to make synthesis gas and, after gasification, to increase the hydrogen content of the synthesis gas before it enters the catalytic reactor. For CTL plants employing extensive carbon management (more than 85-percent capture of plant-site emissions), additional water will be consumed to remove the carbon from fuel gases used for power generation. This water demand is inherent in the chemistry of producing liquids from coal. Stoichiometric analysis shows that the theoretical minimum water requirement is slightly less than one barrel of water per barrel of FT liquid fuels.

In practice, CTL plants will consume additional water for essential plant operations. Water is used in certain in-plant processes and subsystems and plays an inherent role in heat transfer and in the cogeneration of electric power. While these in-plant and plant-utility uses generally involve extensive water recycling, some fraction has to be removed to avoid the buildup of contaminants that would be harmful to processes or equipment. Considering water consumption to meet both stoichiometric requirements and essential processing needs, we estimate the practical lower bound for water usage at a CTL plant at less than 1.5 barrels per barrel of FT or MTG liquid fuels.

The amount of water consumed by a CTL plant will depend on the availability of suitable water supplies, including groundwater. Where water supplies are abundant,

---

[11] In 1972, a breach of the Buffalo Creek coal-slurry impoundment in a West Virginia mine owned by the Pittston Corporation killed 125 people. In 2000, a breach of an impoundment in Kentucky owned by Martin County Coal released 300 million gallons of slurry into the surrounding area (Frazier, 2003).

as they are in certain locations in the eastern and central regions of the United States, plant designs will probably incorporate water-based cooling systems, such as evaporative cooling towers, and involve less recycling of process and utility water. Most of the published conceptual designs for FT CTL plants assume that water is abundant and inexpensive; therefore, the designs provide for extensive water consumption, because that approach minimizes investment and operational costs. For a CTL plant using Illinois bituminous coal, one estimate places water consumption at between eight and ten barrels per barrel of fuel product (SSEB, 2006, Appendix D). Another recent design (NETL, 2007b) for a plant with abundant water supplies yields a consumption rate of five to seven barrels per barrel of fuel product.[12]

In contrast, plants built in arid regions, such as the coal-rich areas of Wyoming and Montana, will likely employ methods to minimize the consumption of surface waters. How much less consumption will depend on cost-benefit and regulatory analyses that would be done as part of a front-end engineering design. Technically, the means are available to reduce water consumption to levels that approach the practical limits just described.[13] Incorporating dry cooling towers into the design will dramatically reduce water consumption, although, based on the currently available low-definition designs, the cost implications of this strategy are highly uncertain. It may also be possible to find sources of water that are unsuitable for other purposes. For example, both subbituminous coals and lignite contain high levels of water that could be used in gasification or other plant processes. Another option could be upgrading nonpotable water. Front-end engineering design studies for early CTL plants will likely include analyses of alternative water-management options. If the results of these analyses are made public, much better information on water consumption and the costs of alternative water-consumption strategies will be available.

The enormous coal resources of Montana and Wyoming are capable of supporting at least one million bpd of CTL production, although, as discussed in Chapter Three, reaching that level will take a couple of decades. At the one million bpd production level, a CTL industry will require the consumption of at least 1.5 million bpd of water. At present, it is difficult to predict how future, more technically mature CTL plants would manage water supply and consumption. While water consumption may be a limiting factor in locating CTL plants in arid areas, at present, this remains an unresolved issue. If and when industrial interest in CTL development grows to the point at which multiple large plants are planned in arid regions, local, state, tribal, and federal governments should assess how long-term water supplies and projected demand

---

[12] This information was obtained in a personal communication with Michael Reed of the National Energy Technology Laboratory (2007).

[13] A design analysis of a CTL plant that would be located in Wyoming and use subbituminous coal predicts a water-consumption rate of one barrel per barrel of product (Gray, Salerno, and Tomlinson, 2005). As Gray stated (SSEB, 2006, Appendix D), "To perform a detailed water balance throughout these plants is complicated and a more thorough analysis would be necessary to have more confidence in these water-use estimates."

will be affected. Otherwise, heavy water usage in early CTL plants will compete with other priority uses and possibly foreclose further CTL development.

Impediments to Private-Sector Investment

Where Are the Major Companies?

The firms most capable of overseeing the design, construction, and operation of multibillion-dollar CTL plants are the major petrochemical companies, such as ExxonMobil, Chevron, ConocoPhillips, and Shell. Each of these firms has the technical capabilities and the financial and management experience necessary for investing in, building, and operating multibillion-dollar projects. Yet none has announced interest in building and operating a first-of-a-kind CTL plant in the United States. Informal discussions with representatives of some of these companies suggest that interest in CTL development does exist within their organizations but that corporate investments are currently being channeled at petroleum-supply opportunities that offer attractive rates of return at much lower oil prices than the $55 to $65 range required by coal-derived liquids.

A few large high-technology firms have indicated their willingness to participate in early CTL commercial development by licensing their proprietary technologies. Notable examples include ExxonMobil, Shell, and General Electric. Such agreements can be structured so that they provide important technical assistance to the smaller firms engaged in building early CTL plants.

If the U.S. government decides to promote CTL development, consideration should be given to obtaining a better understanding of the types of policy measures that might encourage greater participation—including equity investments—by major oil companies and large chemical firms, such as Dow Chemical and DuPont. These and similar firms are most likely to view investment in first-of-a-kind CTL plants as a long-term business-development opportunity. Because of their technical and management strengths, such firms are well postured to exploit the learning that would accompany early production experience and to apply that knowledge in the construction and operation of future plants.

Uncertain World Oil Prices

Discussions with proponents of CTL development indicate that three major uncertainties are impeding investments in CTL production facilities:

- uncertainty about CTL production costs (reviewed in Chapter Three)
- uncertainty regarding how and whether to control greenhouse-gas emissions (reviewed in both Chapter Three and this chapter)
- uncertainty regarding the future course of world crude oil prices.

Of these three factors, the greatest impediment appears to be uncertainty regarding the future course of world oil prices. For CTL to attract private investment, the minimum average price (benchmarked to West Texas Intermediate) that crude oil must sustain over the operating life of the CTL plant is between $55 and $65 per barrel, as discussed in Chapter Three. While this price range is well below recent prices quoted for West Texas Intermediate, considerable uncertainty remains, especially in the investment community, as to whether oil prices will remain above these levels over the next few decades.

Many oil-company executives and industry analysts believe that forecasting future oil prices is a fool's game. On the one hand, oil prices could remain high, because of depletion of large reservoirs, the growing international trend of national control of oil production, and increasing petroleum demand, especially from the developing economies of Asia. On the other hand, oil prices may fall, because of continuing technology advances, greatly increased investments in exploration and production in response to the rapid run-up in crude oil prices since 2000, and the possibility of controls on greenhouse-gas emissions. Complicating any assessment of longer-term trends is oil-price volatility, a problem that has been endemic to the oil industry since its beginnings (Maugeri, 2006; Yergin, 1991).

Figure 6.1 illustrates the sensitivity of the anticipated internal rate of return (post-tax, real) both to world oil prices (benchmarked to West Texas Intermediate) and to

Figure 6.1
Internal Rate of Return (post-tax, real) Versus Crude Oil Prices (West Texas Intermediate)

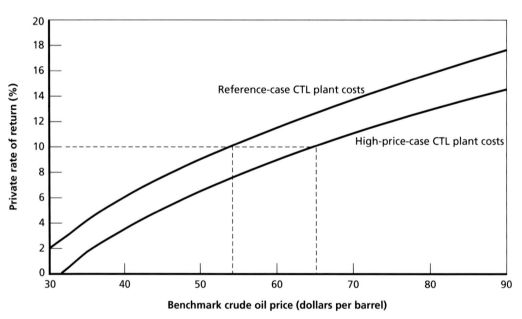

assumptions regarding CTL production costs. The upper curve corresponds to a CTL plant built using our reference case, which represents the low side of our estimate for the construction and operating costs of a first-of-a-kind FT CTL plant (see Appendix A). The lower curve corresponds to our high-plant-cost case (Appendix A). As shown, the rate of return drops to unacceptably low levels—6 and 4 percent for the reference and high-plant-cost cases, respectively—if world oil prices drop to $40 per barrel over the life of the project.

Other factors increase the anticipated range of outcomes. For example, lower rates of return would result if costs are incurred for carbon dioxide emissions, through either a straight tax or an emission credit–trading program, or if, in response to such costs, the carbon dioxide is sequestered. A few firms have announced plans to build FT CTL plants producing less than 10,000 bpd of liquid fuels. Diseconomies of scale suggest that such small plants could have production costs considerably higher than those in the high-cost case described in Appendix A. Some of these firms are also planning to use technologies that have not seen commercial operation at the planned scale of their proposed CTL plant, thereby opening the door to increased construction costs and performance shortfalls (Merrow, Phillips, and Myers, 1981). However, not all uncertainties lead to negative outcomes. Higher anticipated rates of return would result if CTL plants were able to sell captured carbon dioxide for use in enhanced oil recovery or in government-sponsored demonstrations of carbon sequestration.

CHAPTER SEVEN

Designing Incentives to Encourage Private Investment

What is the best way for the government to encourage private investors to pursue early CTL production experience? A variety of government financial instruments have been introduced and discussed for the purpose of promoting early commercial development of advanced energy technologies, including CTL. For purposes of policy analysis, each of these government incentives can be viewed as falling into one of the following five categories:

1. A purchase guarantee with a preset purchase quantity and fixed price for CTL fuel (e.g., the government could contractually agree to purchase all or a portion of the fuel from a CTL plant at a predetermined price)
2. A price floor for a preset amount of the CTL fuel (e.g., the government could agree to purchase the CTL fuel at market prices or at the price floor, whichever is higher)
3. Subsidies that reduce the private firm's investment cost, including investment-tax credits, accelerated tax depreciation, or direct payments of a portion of construction costs
4. Subsidies that reduce the private firm's operating cost or increase revenues, such as the tax credit provided to ethanol blenders
5. A government loan or loan guarantee for a portion of the private firm's debt financing.

In addition to these incentives, the government can also seek an agreement with the private firm to share net income when oil prices are high. This instrument provides a means by which the government can be compensated for the costs and risks associated with implementing any of the preceding five instruments. Approaches for income sharing include establishing a ceiling price for government purchases of CTL fuel or providing the government with an increasing share of the profits made by a CTL plant as the price of petroleum rises above a threshold level.[1]

---

[1] This arrangement is analogous to the pricing terms found in many oil-production contracts outside the United States.

Because our research indicates that major uncertainties are currently preventing private-sector investment in commercial CTL plants, the first step in designing financial incentives is to identify approaches, using these instruments, that eliminate or significantly reduce the probability of outcomes that are so adverse as to preclude private investment. We consider policies that do this over a broad range of potential futures to be *robust*.

We apply two complementary analytic approaches to assess the desirability of alternative incentives. The first approach draws from the empirical literature on contract design and considers any incentive package as a formal relationship between a principal (in our case, the government) and an agent (a specific private investor) that the government seeks to induce to pursue the government's goals—namely, pursue and obtain early experience in CTL production. The second analytic approach considers, in the context of several possible future scenarios, the cash flows that would result for both the government and the private investor if one or more incentives were put in place. It calculates both the real after-tax internal rate of return that a private investor could expect from investing and the real net present value of cash flows to and from the government.[2] Applying these tools yields four key findings that we report in this chapter along with associated policy implications.[3]

Designing an Effective Long-Term Public-Private Relationship

Empirical evidence on the design of agreements between buyers and sellers (or principals and agents) that survive over time in competitive markets can help us anticipate what kinds of public policies would be most cost-effective in a CTL incentive program. Available evidence suggests that such agreements tend to have the following five characteristics.[4]

Relative Control

The more control any party to the agreement has over the execution of the agreement, the more responsibility—and, therefore, risk—the agreement assigns to that party, and vice versa. This limits the potential for moral hazard in a relationship, which occurs when one party's pursuit of its own interests injures another party.

---

[2] Net present value is a standard measure used to evaluate multiyear costs and benefits by discounting future cash flows.

[3] This chapter includes highlights of analytic and policy issues that are addressed in more detail in Camm, Bartis, and Bushman (2008).

[4] See especially Masten (2000), Goldberg (1989), and Rubin (1990). Camm, Bartis, and Bushman (2008) provide additional useful references.

Application. For CTL projects that are being considered for government incentives, this implies that, all else being equal, the more control the investor has over project design and execution, the more responsibility and risk should shift to the investor. And the smaller the investor's stake in the project, the greater the need for the government to assume responsibility for due diligence and focused project oversight. In particular, when selecting an investor, the government should take great care to avoid adverse selection—i.e., the selection of an investor that is unlikely to achieve the government's goals.

Relative Risk Aversion

The more risk averse any specific party to an agreement is, the less risk the agreement assigns to that party.

Application. The federal government is better able than private investors to tolerate the potential losses associated with a failed multibillion-dollar CTL project.[5] Thus, if early CTL production experience is a government priority, an incentive package should shift risk from the investor to the government. And the smaller the investor drawn to the project, the more risk the government should expect to bear. This form of insurance offers a common way for a risk-averse organization to shift risk to another entity that is better able to bear it. For CTL projects, the government can provide such insurance by limiting an investor's downside exposure by establishing a price floor, for example. Income sharing when prices are high can be viewed as a means of compensating the government for providing this insurance.

Cost of Relationship

If the parties to an agreement can administer it in ways that reduce the administrative costs without reducing the level of mutual benefits, the agreement should take advantage of such opportunities.

Application. All else being equal, the government should favor policy instruments that are easier to administer—for example, subsidy mechanisms that can be administered through the existing tax infrastructure.

Relative Cost Advantages

If any specific party to an agreement has any special cost advantages relative to other parties, the agreement should take advantage of them where possible. Current OMB (1992) policy sets the government discount rate for public investment at 7 percent in

---

[5] In general, in cost-benefit analysis, the federal government views risk from a social perspective—i.e., from the perspective of the economy as a whole. The size and breadth of the economy's investment portfolio relative to any private party and the absence of any serious risk of bankruptcy for the economy as a whole make it likely that the government will be less risk averse than private investors are. OMB (1992) policy reflects this perspective.

real terms after adjusting for inflation.[6] This discount rate is significantly lower than the pretax, real private cost of capital typically used to assess the value of an investment. Because the government discount rate is lower, the government places much higher value on out-year cash flows than do private investors.

Application. The larger the real, after-tax difference between government and private discount rates, the more costs the government should accept early in the project relative to those assumed by the investor.

Preservation of Relationship

Agreements that can benefit all parties should seek to sustain themselves by encouraging those parties to remain in the agreements.

Application. This favors contractual terms that respond to unexpected changes in costs, prices, performance, and so on, especially those changes that might encourage the investor to withdraw before sufficient early CTL production experience has been accumulated.

Summary

These five characteristics should shape the behavior of the government and the investor as they approach and then execute a CTL production project. Taken individually, the five characteristics often point in different directions. The challenge in policy design is to seek approaches that apply all factors in a balanced way.

Assessing Financial Effects Under Conditions of Uncertainty

When any CTL incentive package designed to reflect the five characteristics of sound agreements described in the preceding section is applied to a specific CTL project, we can project the cash flows that the package will generate for the government and the investor over the project's lifetime. Cash-flow analysis can transform data on these cash flows into measures of financial performance relevant to the government and investor. This section describes our approach to this analysis.

Basic CTL Project Design

Our cash-flow analysis focuses on investment in and operation of a CTL plant that produces about 30,000 bpd of liquid fuels. We assume investment and operating costs, as described in Appendix A. Investment occurs for five years; production then follows for 30 years. Once the plant reaches its full liquid-fuel output, production continues at

---

[6] Specifically, we refer to the discount rate to be used in "benefit-cost analyses of public investments and regulatory programs that provide benefits and costs to the general public" (OMB, 1992, p. 7).

that level. We also assume that input and product prices remain constant, in real terms, over the life of the project.

Performance Measures and Approach

We use two performance metrics: real after-tax internal rate of return for a private investor and real net present value of cash flows to and from the government. Using a real internal rate of return removes inflation from our analysis.[7] We state real net present value in first-quarter 2007 dollars.

To measure the cost-effectiveness to the government of alternative CTL incentive packages, we calculate the cost to the government of increasing the real private after-tax internal rate of return by one percentage point. To assess the value of this cost-effectiveness metric for any policy change, we introduce the policy change, measure how the change alters the values of private after-tax internal rate of return and government net present value, and divide the change in government net present value by the change in private after-tax internal rate of return. This ratio represents the cost to the government of increasing the real private after-tax internal rate of return by one percentage point.

In our analysis, we do not attempt to identify a hurdle internal rate of return— the minimum level of internal rate of return that a private investor requires to justify the investment.[8] Instead, we focus on packages of incentives that tend to yield a real, after-tax internal rate of return in the range of 5 to 15 percent no matter what world oil prices turn out to be. Within this range, we are most interested in asking how much it costs the government to raise private after-tax internal rate of return by one percentage point.

The analysis uses OMB's prescribed 7-percent real discount rate to calculate how cash flows to and from the government affect the real net present value that the government associates with the project. The cost to the government rises as this net present value falls. In our search for policies, we give more attention to robustness in private internal rate of return than in government net present value, consistent with the government's greater ability to accept risk. With regard to government costs, we favor the central tendency of government net present values over extreme values when examin-

---

[7]   The nominal internal rate of return, d_N, and the real internal rate of return, d_R, are related by the following formula: $d_N = d_R + i + id_R$, where i is the inflation rate.

[8]   In fact, different investors will likely have very different hurdle rates. For example, some firms may value the information gained from building and operating an early CTL plant as a means of pursuing cost improvements and establishing a foundation for future investments. These firms may be willing to accept an internal rate of return well below any hurdle rate normally applied to investments in industrial plants. The government could potentially benefit from identifying such companies and favoring them in any process used to select investors for government assistance.

ing any one incentive package. In this approach, the government, in assessing its potential costs, would give greater weight to more likely scenarios of the future.[9]

Project Financing

The effects of various incentive packages depend on what form of financing a private investor uses. In this chapter, our analytic examples address incentive packages under the assumption of 100-percent equity financing. In addition, this chapter includes a summary of key findings from our companion report (Camm, Bartis, and Bushman, 2008) that explicitly addresses situations in which a substantial portion of the project is financed through debt.

Addressing Uncertainty

As discussed in previous chapters, our research examined the effects of each incentive package over a range of future oil prices, assumptions regarding CTL project costs, and carbon dioxide–sale prices or disposal costs. The analysis presented in this chapter focuses on the first two parameters:[10]

- Crude oil price: We hold the real price of crude oil, benchmarked to West Texas Intermediate, constant over the 30-year operating period of the plant. We examined per-barrel prices that lie within the range from $30 to $90. The low end of this range is consistent with recent low-oil-price projections (EIA, 2007a). At world oil prices above $90 per barrel, the policy implications of government subsidies to early production experience are straightforward; both government and private investors would reap large benefits. (See Figure 6.1 in Chapter Six and Figure 7.1 in this chapter).
- Project cost: We consider both the reference case and a high-plant-cost case, as defined in Appendix A and discussed in Chapter Three. For analytic purposes, we assumed that a front-end engineering design study would be conducted to resolve this uncertainty before an incentive package would be negotiated.

---

[9] OMB's formal policy on how to treat uncertainty about the future is unclear with respect to both when a specific cost-benefit analysis supports government investment and how OMB should "score" such an investment in the budgeting process. The score would determine what cost OMB would use to assess whether the government has the obligational authority to make the investment. Camm, Bartis, and Bushman (2008) go into more detail on these issues.

[10] The results being described in this chapter assume capture of carbon dioxide but no costs (or income) from its disposal (or sale). This case could apply to a situation in which the costs of delivering carbon dioxide to an enhanced oil recovery project are just balanced by the fee paid by the user. In the companion report (Camm, Bartis, and Bushman, 2008), we address potential profits from carbon dioxide sales and potential disposal costs.

Findings and Policy Implications

Uncertainty Prevails About How Any Policy Package Would Affect Private Investors and the Government

No matter which policy package we examine, we find that its effects on investors and the government vary dramatically across potential futures, especially across a range of future oil prices. In Chapter Six, Figure 6.1 illustrated this point from an investor's perspective for the baseline case in which no special government incentives are provided and the CTL project is 100-percent equity-financed. Figure 7.1 covers the same baseline case across the same range of oil prices but shows the effect on both private investors (via private internal rate of return) and the government (via public net present value). Note that Figure 7.1 uses the reference case for CTL project costs, allowing variation only in average oil prices to characterize uncertainty about the future.[11]

Figure 7.1
The Baseline Case: Private and Government Effects with No Incentives in Place

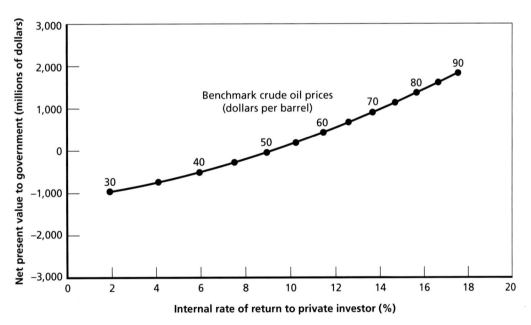

NOTE: This graph illustrates private and government effects as real world oil prices (benchmarked to West Texas Intermediate in 2007 dollars) vary from $30 to $90 per barrel. The graph is based on the reference-case CTL costs, but otherwise, the same assumptions apply that underlie Figure 6.1 in Chapter Six: no special government incentives and no costs or income from disposal of captured carbon dioxide.

RAND MG754-7.1

---

[11] The high degree of variation shown in Figure 7.1 would increase further if we included additional sources of uncertainty, such as potential penalties for emitting carbon dioxide into the atmosphere, potential construction-cost overruns, or delays in plant start-up.

Figure 7.1 indicates real private after-tax internal rate of return on the horizontal axis and real net present value to the government on the vertical axis. Each point in the figure displays these two effects at one average real price for oil over 30 years of CTL production from the project. Real private after-tax internal rate of return varies from about 2 percent if the average oil price is $30 per barrel to almost 18 percent if the average oil price is $90 per barrel. The real net present value of cash flows to the government varies from –$1.0 billion at a petroleum price of $30 per barrel to $1.8 billion at a petroleum price of $90 per barrel. These cash flows reflect the entire effect of the project on government tax revenues.[12]

This degree of variation is large relative to the effect of any change in public policy one might consider. Since the purpose of the incentive package is to cause a private investor to pursue early experience with CTL production, the objective of our analysis is to identify incentive packages that raise the private internal rate of return when it would otherwise be low, as in Figure 7.1, when oil prices are low, but that incur no or very low government costs when government encouragement is not required. For any given cost to the government, an incentive package is *preferred* the more it increases private internal rate of return when the internal rate of return is too low to encourage private-investor participation in a project.

A Policy Package of a Price Floor, Income Sharing, and Investment Subsidies Can Be Both Robust and Cost-Effective

In this section, we show how the government can combine individual policy instruments to form preferred policy packages. As we compared alternative financial-policy packages, our analysis found that combinations of the following three instruments were particularly attractive:

- Price floors for CTL-based fuel allow targeted efforts to raise real private after-tax internal rate of return in futures in which average oil prices are low. Price floors do this particularly well because they impose no costs on the government for futures in which average oil prices alone are high enough to promote investor participation.
- Income sharing provides a mechanism that the government can use to recover the public costs associated with promoting early investment, if and when oil prices are high. In exchange for a government-provided price floor or other investment incentives that limit the investor's downside risk, the investor would pay the government a portion of its net income when its income, after paying taxes and the government's portion of the sharing agreement, is large enough to ensure the continued viability of the project.

---

[12] Negative net present values result because the income tax on profits from CTL sales cannot offset the loss of tax revenues from depreciation of the capital invested in the plant.

- Investment subsidies, such as investment-tax credits and accelerated depreciation, are cost-effective instruments across many potential futures. They are a good option for raising real private after-tax internal rate of return no matter what future occurs. This result reflects systematic differences in government and private costs of capital.

To illustrate how these instruments can be combined, we begin with an example that combines just the first two—namely, a policy package that combines a price floor of $40 per barrel of CTL-based fuel with a net income–sharing agreement that would occur after average world oil prices reach $60 per barrel. Figure 7.2 shows how this package affects the investor and the government relative to the baseline case depicted in Figure 7.1. Each arrow shows how the package moves the private real after-tax internal rate of return and real government net present value at one future average oil price from the baseline case (circles) to the policy package (diamonds). This package has no effect when oil prices are between $40 and $60 per barrel, since the incentives take effect only when oil prices are outside that range. The package significantly enhances private internal rate of return at average oil prices below $40 per barrel at the expense of the government. It also improves the government net present value at future average

Figure 7.2
Policy Package A: Effects of Introducing a Price Floor and a Net Income–Sharing Agreement

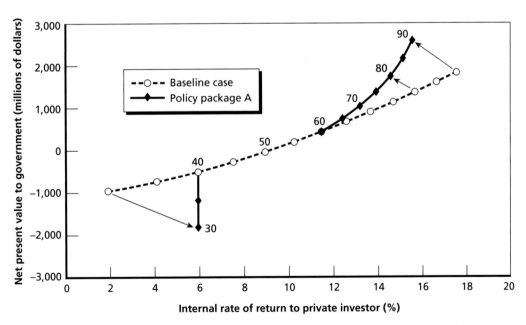

NOTE: Based on reference-case CTL costs, $40 per barrel price floor, income sharing above $60 per barrel, and no income or expenses for disposing of captured carbon dioxide.
RAND *MG754-7.2*

oil prices of $60 or higher at the expense of the investor. For this particular example, income sharing is achieved by linearly raising the combined federal and state effective tax rate from the baseline case of 36.11 percent at oil prices of $60 and lower per barrel to 51 percent for a future in which the average world oil prices would be $90 or more per barrel (Camm, Bartis, and Bushman, 2008). This particular formula for income sharing ensures that, while the government benefits more at higher average oil prices, the investor's after-tax internal rate of return also continues to rise as oil prices increase. This serves to preserve the relationship between the two parties.

Figure 7.2 captures the essence of the insurance arrangement represented by this package. The government indemnifies the investor against loss if future average oil prices are low, in exchange for substantial payments if future average oil prices are high. The effect is to narrow the range of outcomes across alternative futures for the investor and to allow the government's range of outcomes to increase. This can be appropriate when the government is less risk averse than the investor.

Would the package shown in Figure 7.2 be sufficient to motivate early private investment in CTL plants? We cannot answer that question because we do not know the nature of competing investment opportunities. If investor interest requires a real after-tax internal rate of return of 10 to 12 percent (equivalent to a nominal pretax internal rate of return of 20 to 23 percent), the package shown in Figure 7.2 will be acceptable only if investors are convinced that oil prices are very likely to stay above $55 (in 2007 dollars) per barrel over the operating life of the CTL plant. The price floor of $40 guarantees that the investor will realize no less than a 6-percent real after-tax internal rate of return.[13] Yielding the equivalent of a nominal pretax internal rate of return of 13 percent, the price floor provides the investor with a respectable rate of return in the event that oil prices drop well below expectations during the operating life of the CTL plant.

Figure 7.3 shows the effects of moving to an alternative policy package—policy package B—that also is robust. Here, the price floor increases from $40 to $45 per barrel of CTL-based fuel, and a 10-percent investment tax credit is added. The income-sharing formula remains unchanged. The arrows now show the effects of moving from policy package A described in Figure 7.2 (diamonds) to this new policy package B (triangles).

Three points are important here. First, in all futures, policy package B clearly enhances private real pretax internal rate of return more than the first package but at the expense of the government. This is immediately apparent because all arrows point to the right and down.

---

[13] Note that achieving this minimum internal rate of return requires that the investor or owner bring the CTL plant to production within budget (reference-case costs) and on schedule and that the carbon dioxide captured at the plant can be disposed of at no further costs to the plant owners.

Figure 7.3
Policy Package B: Effects of Raising a Price Floor and Adding an Investment-Tax Credit

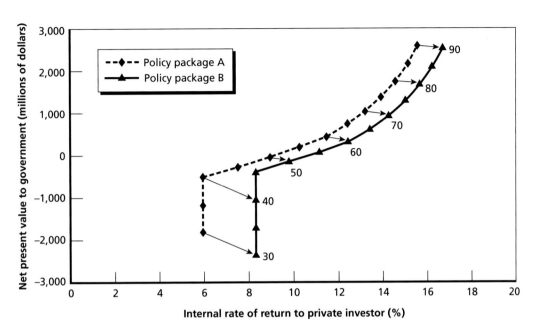

NOTE: Based on reference-case CTL costs, changing price floor from $40 per barrel to $45 per barrel, adding a 10-percent investment-tax credit, and income sharing above $60 per barrel (same as Figure 7.2), and no income or expenses for disposing of captured carbon dioxide.
RAND MG754-7.3

Second, despite the significant shift in internal rate of return and net present value, the new package still displays the basic pattern of an insurance arrangement as was associated with the package shown in Figure 7.2. The range of outcomes for the investor is narrower, and the range of outcomes for the government is broader than in the baseline policy case.

Third, we can decompose the shift from package A to package B into two changes—the increase in the price floor and the addition of an investment-tax credit. Arrows above $45 per barrel show us the effect of adding the tax credit; those below show the joint effects of both at each future average oil price. The arrows to the right of $45 per barrel are not as steep as those to the left, which indicates that, at any average oil price, the cost to the government of increasing real private after-tax internal rate of return by one percentage point is lower when the government adds an investment-tax credit than when it increases the price floor on CTL-based fuels.

Which instrument—a tax credit or a price floor—is better? At any future average oil price, the tax credit helps the investor and provides that help sooner—namely, during the construction stage of the project—than does the price floor. The difference in government and investor costs of capital makes the investment-tax credit more cost-effective at any particular future average oil price. This demonstrates the relative desirability of an investment subsidy at all future average oil prices. But a price floor, although a less cost-effective way to raise private internal rate of return at any average oil price, imposes costs on the government only in the presumably less probable event that oil prices would be very low when investors are in greatest need of protection. When combined in a policy package, they complement one another.

The policy packages shown in Figures 7.2 and 7.3 are based on specific examples for the investment subsidy and income sharing. Effective alternatives can be constructed using accelerated depreciation or direct government investments in place of or in addition to an investment-tax credit. Likewise, there are many ways to craft an equitable income-sharing agreement. That said, in a qualitative sense, such alternative packages are likely to look very much like the balanced packages shown here.

If Better Information on Project Cost Is Available, Policy-Package Design Can Be More Balanced

The analysis in the previous section assumes that the reference case holds for CTL plant construction and operating costs. Suppose instead the case in which CTL plant costs are at the high end of the estimates established in Chapter Three. How would the preferred policy packages be altered to reflect higher plant construction and operating costs? What difference would the change make for investors and the government? Answers to these questions illustrate the value of waiting to finalize the government's policy package for encouraging early CTL production experience. It is crucial to know as much as possible about how the project itself will financially perform before construction of the production facility even begins.

Our analysis identified a variety of policy packages of this kind when the high-cost case holds for CTL project costs. One package combines a price floor of $45 per barrel of CTL-based fuel, a 10-percent investment-tax credit, 100-percent expensing of investment costs in the first year of production for tax purposes,[14] and the same net income–sharing agreement described already. Figure 7.4 shows how this package (diamonds)—policy package C—affects the investor and the government relative to the baseline of no incentives (circles) that applies when the high CTL plant cost case holds. Arrows show how real private after-tax internal rate of return and real government net present value change when this package is introduced.

By design, this new package yields a range of outcomes for the private investor that is similar to that shown for policy package A, illustrated in Figure 7.2. For example, the minimum internal rate of return is only slightly above 6 percent, and crude oil prices need to be $57 and $65 per barrel to obtain an internal rate of return in the 10- to 12-percent range. This package now affects private internal rate of return and

---

[14] Accelerated depreciation allows the taxpayer, when calculating taxable income, to write off investment costs against revenues earlier in the asset lifetime. The most extreme form of accelerated depreciation—100-percent expensing—allows the taxpayer, when calculating taxable income, to write off all costs of investing in an asset in the first year in which it produces income.

Figure 7.4
Policy Package C: Effects of a Robust Policy Package Designed for the High-Cost Case

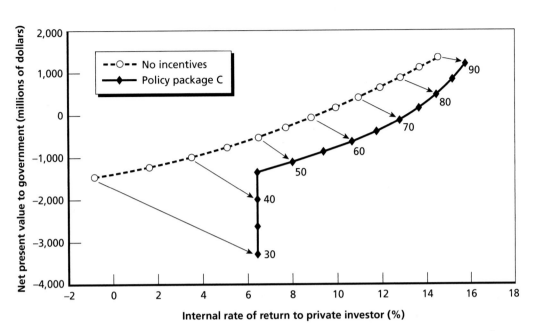

NOTE: Based on high CTL plant costs, a $45 per barrel price floor, a 10-percent investment-tax credit, accelerated depreciation, income sharing above $60 per barrel, and no costs or income from disposal of captured carbon dioxide.
RAND MG754-7.4

government net present value at all average oil prices because more aggressive incentives are required to keep private internal rate of return within our target range. Project costs are now significantly higher than before; therefore, the costs to the government are higher. For example, in the event that oil prices drop to as low as $30 per barrel, government costs for this case could be very high—as much as $3.3 billion in net present value. If the government believes that there is a significant probability that such low oil prices might prevail, it may decide to limit significantly the amount of early CTL production that it will promote.

To offset higher project costs, the robust packages of policies we identified in policy package C tended to use higher price floors and more aggressive investment subsidies, together with net-income sharing, than the packages we identified in the reference case. This showed us that the structure of a balanced policy package can change in subtle ways when expectations about project costs change. Before finalizing the design of a policy package, it pays to wait for the best information available on project cost and performance.

Loan Guarantees Can Have Powerful Effects but Require Great Caution

Another powerful policy instrument the government could use to encourage CTL investment is loan guarantees, which facilitate debt financing. Debt financing allows an investor in a project to increase its real after-tax internal rate of return as long as the cost of debt capital available to the investor is below the internal rate of return for the cash flows generated by the project. An investor can increase its real after-tax internal rate of return by increasing the debt share as long as a higher debt share does not excessively increase the risk of default. Risk of default always increases with rising debt share because the cost of making payments to the lender can exceed a project's realized cash flow.[15] Recognizing this, lenders generally expect higher default rates for projects financed with higher debt shares. All other things being equal,[16] lenders respond by charging higher interest rates when investors maintain higher debt shares, and they ultimately limit how much debt capital they will offer for a project. Given how lenders behave, an investor chooses its debt share to balance the advantage of having more debt capital available with the disadvantage of incurring a higher risk of default.

To the extent that a risk of default exists, a government loan guarantee allows a lender to offer loans to an investor at a lower rate by shifting the risk associated with default from the lender to the government. If no default risk exists, a loan guarantee will have no effect. The larger the default risk, the more a loan guarantee will reduce the rate a lender offers the investor. A loan guarantee becomes a more powerful investment incentive precisely when it imposes a larger expected cost on the government agency offering the guarantee. In this regard, the guarantee acts very much like a direct government subsidy. But it is less visible than a direct subsidy, because the government incurs a cost only if default occurs and, one would hope, the probability of a default is less than one.

A loan guarantee can encourage an investor to pursue early CTL production experience in two ways. First, if the investor maintains its debt share, the guarantee reduces the interest rate on the loan. By definition, this reduces negative cash flows over the life of the CTL project for the investor and increases its real after-tax internal rate of return.

Second, a loan guarantee encourages an investor to increase its debt share. Because the loan guarantee lowers the risk of default, the lender is less likely to charge more for loans or to limit the size of a loan when the investor seeks to increase its debt share. Because the investor's cost of debt capital does not rise as much when it seeks a higher debt share, the investor chooses a higher debt share if it can. This tendency to expand

---

[15] Without debt, a project defaults when operating costs exceed revenues. With debt, default occurs when the sum of operating costs and debt payments exceeds revenues.

[16] Many factors could condition a lender's response, including the reputation and past performance of the borrower and its suppliers, partners, and customers; the current state of the economy or economic sector in which the investment lies; and beliefs about the inherent technical riskiness of the project at hand.

debt share exists precisely because of the government's willingness to bear a portion of the risk of default. But by creating an incentive for the investor to increase its debt share, the government also increases the probability of default and thereby the potential cost to the government of offering the loan guarantee.

In effect, as a loan guarantee shifts more risk from a lender to the government, the government agency should assume more of the fiduciary responsibility that the lender had to its shareholders. That is because, as the guarantee shifts risk from lender to government, it is really shifting risk from shareholders to taxpayers. To assume this new fiduciary responsibility, the government agency must perform due diligence—verify the financial, managerial, and technical capacities of the borrower—and monitor the borrower's performance through the course of the loan in the same way that the lender would have if it still had such fiduciary responsibilities.

The quantitative size of the effects described above—i.e., how much an investor would increase its debt share in response to a loan guarantee or how much that increase in debt share would increase the cost of the loan guarantee to the government—varies from one project to the next. We did not explore uncertainties associated with project performance that would allow us to estimate the size of these effects for the hypothetical CTL project examined here. However, our analysis does yield three major findings:

- *Except at very low expected petroleum prices, if the investor holds its debt share constant, a loan guarantee has only small effects on real after-tax internal rate of return flows.* Its effects on real after-tax internal rate of return grow rapidly as the loan guarantee induces the investor to increase its debt share more. In the end, a loan guarantee is likely to encourage investor participation in a project mainly by allowing it to benefit from the effects of using a higher debt share, not through its direct effects on the cost of debt capital (see Camm, Bartis, and Bushman, 2008, for details).
- *How much a loan guarantee costs the government depends fundamentally on how much responsibility the government takes to oversee the project.* To limit the potential for moral hazard, when the government accepts the additional risks of a large loan guarantee, the government must be given additional control and oversight over the project, as would be the case with a private lender with a major stake in the project. Moreover, the more the government limits the investor's total debt share and hence the project's probability of default, the smaller the expected cost to the government (but also the smaller the positive effect of the loan guarantee on real private after-tax internal rate of return).
- *The power of any loan guarantee to promote early CTL investment ultimately lies in how much default risk the government is willing to accept.* As noted, a loan guarantee offers an investor no benefits unless a default risk exists. If a loan guarantee exists, the more the government is willing to subsidize the project by guarantee-

ing a larger fraction of the total investment and thereby accepting higher risks, the greater will be the investors' willingness to participate in the project.

The federal government should use a loan guarantee to promote early CTL production experience only with a strong understanding of how that guarantee works, how much it is likely to cost the government to encourage investor participation to any degree, and how well the government can put in place an effective project monitoring and control system capable of protecting the federal purse. The structure of a loan guarantee and the associated project oversight depends crucially on the investing firm, its overall project approach, and the technical and other risks associated with the project. Lacking this project-specific information, we have no basis for recommending specific parameters for a CTL loan-guarantee program. Additional background on loan guarantees is provided in our companion report (Camm, Bartis, and Bushman, 2008).

Promoting Competition

Different kinds of investors will prefer different packages of incentives. For example, large well-capitalized firms are likely to prefer up-front incentives, such as investment-tax credits and accelerated depreciation, as opposed to either a price floor or a loan guarantee. Because such large firms are better able to bear risks, they may view the prospect, no matter how small, of defaulting on a federal loan guarantee, or of receiving operating subsidies, as politically unacceptable. In contrast, a smaller firm seeking debt capital will likely prefer price floors and loan guarantees because, unlike a larger firm, a smaller firm can bear less risk and face a higher probability of bankruptcy, if a CTL project fails, than does a larger firm, even if both firms have the same expectations for the technical performance of the project.

It is to the government's advantage to have as many qualified firms as possible compete for government incentives. Therefore, it is in the government's interest to offer a menu of incentives, such as the robust policy packages described in this chapter. Firms can select their preferred package, and the government can competitively choose those proposals that are most advantageous to the government, based on a calculation of the likely effects on government net present value in different futures. For example, the government might announce that, for purposes of evaluating offers, it will assign an equal probability for oil prices to be, on average, $45, $65, $80, $90, and $100 per barrel over the operating life of a CTL plant. If a qualified firm proposed policy package C, as illustrated in Figure 7.4, the government net present value would be calculated at each of these five oil prices and the average would be used to compare this offer with that received from other firms, which may propose a very different mix of incentives.

This approach is compatible with the recent federal movement away from writing the statement of work for an acquisition and instead writing a statement of objectives to describe *what* it wants and requiring offerors to write statements of work that describe *how* they will deliver what the government wants. Rolling the package of incentives into a source selection in this way could allow the government to focus on the pattern of net present value outcomes it expects from alternative offers in various futures and not on how to design a package of incentives that would induce this flow. Such an approach is challenging. But by allowing each qualified offeror to frame its own incentive package, it creates opportunities for mutual gains between the government and potential investors with a wide range of interests that a traditional source-selection approach would not allow.[17]

Summary

This chapter describes an analytic way to design a package of incentives that the government can use to promote early investment in CTL plants while limiting the cost to the public treasury in the face of significant uncertainty about the future. The analysis reveals that the incentive packages and implementation policies need to be sensitive to this uncertainty. They should seek to raise real private after-tax internal rate of return in futures when that internal rate of return is too low to induce investor participation, but not in other futures. And they should seek to do this in a cost-effective way, no matter what future occurs.

A balanced package of a price floor, an investment incentive, and an income-sharing agreement is well suited to do this. The investment incentive is a cost-effective way to raise private after-tax internal rate of return in any future. A price floor can cost-effectively provide an additional boost in futures in which oil prices are especially low. And an income-sharing agreement structures the overall incentive package as an insurance agreement between the investor and government, with payments to the government in futures in which oil prices turn out to be high.

Because the most desirable form of a balanced package depends on expectations about project costs, the government should wait to finalize the design of an incentive package until it has the best information on project costs that is available without actually initiating the project. We strongly advise that an incentive agreement not be finalized until both government and investors have the benefit of improved project cost and performance information that would be provided at the completion of a front-end engineering design.

Loan guarantees can strongly encourage private investment. However, they encourage investors to pursue early CTL production experience only by shifting real

---

[17] For more information on this approach, see Camm, Bartis, and Bushman (2008).

CHAPTER EIGHT

Moving Forward with a Coal-to-Liquids Development Effort

The prospects for developing an economically viable CTL industry in the United States are promising, though important uncertainties persist. FT and MTG CTL technologies are ready for initial commercial applications in the United States, production costs appear competitive at world oil prices that have prevailed over the past two years, and proven coal reserves are adequate to support a large CTL industry operating over the next 100 years.

Prevailing Uncertainties

The most important factor presently impeding private-sector investment in CTL production is uncertainty regarding the future course of world oil prices. Our best estimate is that coal-derived liquids are competitive when crude oil prices (benchmarked to West Texas Intermediate crude oil) are between $55 and $65 per barrel. Many investors expect oil prices to remain high over the financial lifetime of a CTL plant. However, all investors are concerned that low oil prices would bring severe adverse consequences—namely, extremely low rates of return or project bankruptcy, with the latter being of special concern to those projects requiring annual repayments of debt.

The most important factor presently impeding federal support of CTL development is the conflict between, on the one hand, CTL development and the consequent increase in coal use and, on the other hand, the growing consensus that the federal government should take measures to reduce greenhouse-gas emissions. Our research has identified technical approaches that might significantly mitigate, if not eliminate, this conflict. However, these approaches are not fully developed. For some—such as carbon sequestration as part of enhanced oil recovery and the combined gasification of coal and biomass—technical risks are fairly low, and economic and environmental viability can be established in a few years. But for others—such as carbon sequestration in deep saline formations—establishing technical, economic, and environmental viability will require multiple, large-scale, long-term demonstrations and require at least a decade of research.

The United States does not have in place a legal and regulatory framework for reducing greenhouse-gas emissions. Over the next few years, we believe that legislation will probably be enacted to establish such a framework. Meanwhile, uncertainty prevails as to when and how the federal government will move to control greenhouse-gas emissions. This issue also extends to the uncertain legal liabilities associated with pioneering efforts to sequester carbon dioxide in the United States. These legal and regulatory uncertainties serve as an additional impediment to investment.

Our research has also revealed an important issue that should be of concern both to potential investors and to the federal government—namely, uncertainty in the cost and performance of first-of-a-kind plants producing coal-derived liquids. Although the FT approach to liquid-fuel production from coal is fully commercial in South Africa, the "newest" South African commercial CTL plant was built during the early 1980s. For the MTG process, the only commercial-scale experience occurred using natural gas in the New Zealand plant that began operations in 1985. A commercial CTL plant has never been built in the United States. If one were to be constructed, it would be based on components, materials, process control, and pollution control and prevention systems that were not available 25 years ago.

At present, the knowledge base for CTL plant construction costs and environmental performance is very limited. All estimates that we could locate are based on highly conceptual engineering-design analyses that are intended to provide only rough estimates of costs. We have learned that, when it comes to cost estimates, typically the less one knows, the more attractive the costs appear to be (Merrow, Phillips, and Myers, 1981). Moreover, engineering analyses incorporating a low level of design detail are unable to address trade-offs between costs and environmental performance, including options to minimize water consumption or to gain access to alternative sources of water, as may be required of plants built in arid regions.

The Military Perspective

The U.S. Air Force and the other services are highly dependent on transportation fuels for maintaining readiness and executing their missions. In 2007, DoD consumed petroleum at an average rate of 330,000 bpd. Jet fuel accounted for about two-thirds of DoD's petroleum use. Fuel oil for naval ships accounted for much of the remainder. With the exception of the ships, nearly all of the mobile and combat-support systems in the U.S. armed forces are fueled by a particular formulation of jet fuel known as JP-8 and its close relative JP-5. These fuels are preferred for combat operations because, compared to gasoline, they have a very high energy density per unit volume and because they are less subject to accidental ignition.

In recent years, long-term DoD interest in pursuing energy conservation has grown more intense within the U.S. Air Force, by far the largest single user of energy in

DoD (Shanker, 2006). DoD's Assured Fuels Initiative of 2001 placed special emphasis on providing the U.S. military with cleaner fuels derived from secure domestic sources, such as coal and natural gas (Harrison, 2006). Because 80 percent of U.S. Air Force energy use involves aviation fuel, the U.S. Air Force has given aviation fuel high priority in its implementation of this initiative. The Energy Policy Act of 2005 (P.L. 109-58) gave DoD several openings to approach energy conservation and unconventional-fuel development more proactively (Barna, 2005).[1]

Surges in world oil prices over the past few years have overwhelmed DoD's budgeting system, leaving DoD—and the U.S. Air Force in particular—with insufficient funding to cover its must-pay fuel costs without cutting programmed spending elsewhere (Dunlap, 2007). In 2005, Hurricane Katrina severely disrupted supplies of refined fuels and drove up their prices briefly but precipitously. This gave the U.S. Air Force a wake-up call about its vulnerability to physical and financial instabilities in fuel markets (Aimone, 2007).

Persistently high oil prices have also made DoD planners very conscious of the possible long-term effects on DoD (Wynne, 2007b). Higher oil prices would reduce what DoD could buy within a fixed top-line budget and so could potentially limit its ability to execute its mission. Only budget increases to reflect higher oil prices would prevent such negative effects. In the face of what they expect to be increasing competition from domestic policy priorities, DoD planners currently expect limited political tolerance for higher defense budgets (adjusted for inflation) over the long term. Furthermore, the military's ongoing high operational tempo in Iraq and Afghanistan has made fuel price and availability issues more important to DoD planners and resource managers than they might have been during a more peaceful period.[2]

In response to rapidly rising fuel prices, the U.S. Air Force has taken a proactive position regarding the development of a commercial FT CTL industry within the United States. Ground and flight testing of blends of FT-derived jet fuel and conventional JP-8 in military aircraft engines are under way, and current plans call for testing and certifying all U.S. Air Force airframes to fly on a synthetic fuel blend by early 2011 (Wynne, 2007c). In support of this certification effort, the Defense Logistics Agency is

---

[1] For example, Section 369 of the act directs the Secretary of Defense, working in coordination with other relevant federal departments, to

> establish a task force to develop a program to coordinate and accelerate the commercial development of strategic unconventional fuels [and] develop a strategy to use fuel produced [in the United States] . . . from coal, oil shale, and tar sands . . . in order to assist in meeting the fuel requirements of the Department of Defense when the Secretary determines that it is in the national interest [and] establish standards for clean fuel produced from domestic sources.

[2] Per Aimone (2007), "Afghanistan and Iraq have increased military fuel use by as much as 56,000 barrels a day. . . . [T]he military uses fuel at twice the rate it did in the first Persian Gulf War and four times the rate it did in the Second World War."

purchasing flight-test quantities (hundreds of thousands of gallons per year over multiple years) of FT-derived jet fuels.

Further, the Air Force is a participant in the Commercial Aviation Alternative Fuels Initiative, which is a forum sponsored by the commercial aviation trade associations and the Federal Aviation Administration.

The U.S. Air Force has also established the goal of being prepared, by 2016, to cost-competitively acquire 50 percent of the Air Force's domestic aviation fuel requirement via an alternative fuel blend in which the alternative component is derived from domestic sources produced in a manner that is environmentally superior than fuels produced from conventional petroleum.[3] If the potential fuel purchases associated with this goal were to be met with 50/50 blends, a production capacity of approximately 20,000 bpd of unblended alternative fuels would be required. Considering the product distribution of an FT CTL plant, attaining this level of jet-fuel production would likely require a domestic FT CTL capacity of between 50,000 and 80,000 barrels (diesel fuel equivalent) per day.

The findings of our study, as presented in Chapter Five, show that CTL development offers major economic and security benefits at the national level. As one of the nation's largest petroleum consumers and as a key component of the nation's defense, the U.S. Air Force has a stake in reaping these nationwide benefits. Beyond these shared nationwide benefits, unique benefits that would accrue just to or primarily to the U.S. Air Force appear to be less significant, especially in light of the risks and uncertainties associated with alternative-fuel development. For these reasons, the policy options for furthering U.S. Air Force plans and efforts regarding CTL commercial production should be developed and implemented in the context of an overall federal policy framework.

Federal Policy Options

With regard to coal-derived liquids, and to unconventional liquid fuels more generally, the federal government can take one of three policy positions. It can adopt a hands-off position and let the marketplace decide whether and how to develop an unconventional-fuel industry. At the other extreme, it can choose to actively promote the development of a commercial unconventional-fuel industry. Based on our research, we suggest consideration of a middle position, which we characterize as an *insurance policy*, in which the government focuses on removing uncertainties and thereby places

---

[3] U.S. Air Force Secretary Michael Wynn focused early thinking on FT-based synthetic fuels (Wynne, 2007a; Billings, 2007). His successor, Secretary Michael Donley, has directed that the alternative-fuel goal be expanded to ensure that all unconventional fuel types are considered, as long as they are produced economically and in a manner consistent with national environmental goals (Aimone, 2008).

both it and the private sector in a better position to respond efficiently as oil markets, regulatory efforts, and technology evolve.

The Hands-Off Position

Since the demise of the Synthetic Fuels Corporation in the early 1980s, the federal government has taken a hands-off position with regard to the development of unconventional liquid fuels derived from such fossil energy resources as oil shale and coal. This position places high value on market processes and private decisionmaking unfettered by government interference. It allows federal support of R&D efforts but not of near-term technology development and demonstration activities. Those favoring this position might argue that, in the past, the federal government has done a very poor job of deciding which energy technologies and projects are best, and thus should stay out of that decisionmaking process.

However, the strongest argument against a hands-off position centers on the existence of benefits that CTL development would yield to the public at large but that private-sector investors would not capture or fully realize. The analysis presented in Chapter Five indicates that appreciable public benefits, or *externalities*, would accompany any measures, including the development of a CTL industry, that would decrease demand for conventional petroleum. Examining only the net domestic economic surplus created by reduced world oil prices, we estimate that each marginal barrel produced generates about $6 to $24 worth of such benefits. As a result, our research suggests that it is cost-effective for the federal government to subsidize CTL production at some level between $6 and $24 per barrel.[4]

Opponents of a hands-off approach can also point to political and national security concerns, including continuing strife and social unrest in many large oil-producing nations, the four-fold increase in world oil prices over the past few years, and OPEC's clear intent to keep oil prices high.

Promoting Commercial Production

Policy options that promote commercial production center on enhancing the competitiveness of fuels from unconventional sources. As such, much of the analysis described in Chapter Seven, which focuses on providing incentives for limited early production, also applies to the broader issue of promoting large-scale commercial production.

Petroleum Tax. One approach is to increase the price of conventional-petroleum products by imposing a federal tax that would not apply to fuels derived from unconventional sources. By raising the cost of petroleum products, this approach has the benefit not only of promoting the commercial development of fuel sources but also

---

[4]   To apply this concept more generally, measures that cause demand for conventional petroleum to drop by one barrel are associated with a premium of $6 to $24 per barrel. A net savings of one barrel from a conservation measure or from the production of a renewable fuel is just as effective as a CTL barrel in reducing world oil prices.

of encouraging energy conservation. Most importantly, this approach is technology neutral and recognizes the value and efficiency of implementation through individual choices by consumers and private industry. Legislation and regulations aimed at implementing this approach would need to deal with the issue of which sources qualify as unconventional. For example, if tar sands are determined to be unconventional fuels, Canadian imports might be exempt from a federal tax under the provisions of the North American Free Trade Agreement. Since the goal of a tax exemption is to reduce use of petroleum, another issue that would need to be resolved is how a tax exemption would treat unconventional fuels that are manufactured with considerable amounts of petroleum or natural gas, as is the case for manufacturing corn-based ethanol.

A tax on petroleum or on selected petroleum products is not a new concept,[5] but, because it involves raising the prices of transportation fuels, many elected officials regard a petroleum tax as politically infeasible. Moreover, even if the federal government were prepared to move forward with a comprehensive solution to reducing petroleum demand, it would likely take several years to phase in a petroleum tax to levels sufficiently high to induce investment in unconventional fuels. Finally, a tax exemption for liquid fuels from unconventional sources could result in significantly greater greenhouse-gas emissions. For this reason, the imposition of a petroleum tax should be accompanied by provisions that would tax or otherwise provide a disincentive to carbon dioxide emissions from the production and use of fossil fuels.

Production Subsidies. An alternative to a tax on conventional-petroleum products is a subsidy for all unconventional products. An example of this approach would be to extend the tax credit that currently supports alcohol fuels to all unconventional fuels. This subsidy approach shares many of the benefits of the previous approach in that it is goal-oriented, technically neutral regarding supply-side alternatives, and deferential to the marketplace. However, it has three major disadvantages. First, the net effect would be to use federal funds to lower fuel prices below what they would otherwise be. As such, this approach discourages energy conservation and increased efficiency in the use of energy. Second, the approach basically represents a fuel credit and, as such, from a cost-benefit perspective, it is among the least efficient incentives available to the government for promoting private-sector investment, as was discussed in Chapter Seven. Third, this policy alternative could adversely impact the federal budget. For example, extending the current corn-ethanol subsidy of $0.51 per gallon of ethanol to the energy equivalent of ten million bpd of gasoline or diesel fuel from unconventional sources would cost the federal government about $116 billion per year.

Other Incentives. Proponents of federal measures have advocated other incentives to promote the development of a large unconventional-fuel industry in the United

---

[5]   For example, in 2006, a task force sponsored by the Council on Foreign Relations recommended consideration of strong policy measures, including a federal tax, to reduce gasoline demand (Council on Foreign Relations, 2006).

States. As shown in Chapter Seven, a combination of incentives is the most cost-effective approach for promoting private-sector investment. But Figures 7.3 and 7.4 demonstrate that, if oil prices drop from their current very high levels, the federal government could incur substantial costs. For example, if oil prices were to average $35 per barrel over the life of the CTL plant, the two figures show that the net present value of government costs associated with a cost-effective incentive package could be more than $1.5 billion (January 2007 dollars) for 30,000 bpd of production. Extending this type of incentive package to strategically significant levels of production could result in government costs having a net present value of between $100 billion and $200 billion.

Chapter Seven also addresses federal loan guarantees. The same issues that were raised regarding individual projects apply to the use of this incentive tool to promote strategically significant levels of production. Federal loan guarantees are a powerful instrument for promoting private-sector investment. But that very power suggests that they must be used with great restraint. Specifically, by transferring risks from investors to the government and by encouraging higher debt/equity ratios, a federal loan guarantee appreciably raises the chances of project default, especially if world oil prices decline significantly. In principle, this problem can be solved by effective government oversight of projects requesting loan guarantees. But the oversight function runs counter to the position that the government should be a catalyst for, as opposed to the arbiter of, economic development by the private sector.

An Insurance Policy

Instead of emphasizing near-term production and risking adverse consequences, we suggest a policy option that recognizes prevailing uncertainties and emphasizes future capabilities (Harlan, 1982). Under this approach, the government would target its efforts at resolving prevailing uncertainties and placing the private sector in a better position to build a commercial CTL industry. The five elements of this strategy are as follows:

- Complete front-end engineering design studies of CTL and dual-feedstock production plants to establish costs, risks, potential economic performance, and environmental impacts.
- Use federal incentives to ensure early commercial production experience with a limited number of first-of-a-kind CTL or dual-feedstock plants to establish performance and provide a foundation for post-production learning.
- Conduct multiple large-scale, long-term demonstrations of the sequestration of carbon dioxide created at electric generation or CTL production plants or both.
- Undertake the research, development, and testing required to establish the technical viability of using a combination of biomass and coal for liquids production.

- Broaden and expand the federal portfolio directed at long-term, high-payoff research relevant to transportation-fuel production.

Cost-Share Front-End Engineering Design

We estimate the cost of completing a site-specific front-end engineering design of a first-of-a-kind commercial FT or MTG CTL plant at between $20 million and $40 million. This estimate includes associated trade-off analyses covering technology choices, environmental controls, and water management. This is a very small fraction of the total construction cost, which would be roughly $1 billion for each 10,000 bpd of capacity.

As part of the insurance-policy approach, we suggest that the government consider cost-sharing a few site-specific front-end engineering designs. This can be done in a way that protects proprietary information while still providing the government, potential investors, and the public with important information regarding the anticipated costs and performance of commercial CTL plants.

We note that a front-end engineering design is a necessary step in progressing toward the final detailed design of a CTL plant. In particular, the front-end engineering design is the basis for obtaining permits at the local, state, and federal levels that will govern construction practices and plant performance requirements and that are a necessary input into the detailed design of the plant. Additionally, we strongly recommend that neither the government nor investors enter into negotiations regarding incentive packages without the more detailed knowledge of potential costs, performance, and risks that a front-end engineering design provides.

Promote Early, but Limited, Operating Experience

In Chapter Seven, we summarized our analysis of incentive packages for promoting early commercial operating experience. Given prevailing uncertainties, government incentives should apply to a very limited number of CTL plants. Based on considerations of coal types and technical options, a reasonable government objective is to have production from three to six moderate-size plants.

The primary reasons for subsidizing early production experience are to facilitate post-production cost improvements and to posture the private sector for a possible rapid expansion of a more economically competitive CTL industry. Accordingly, the government should carefully screen out firms (or collaborations of firms) that are unable to demonstrate the financial, management, and technical wherewithal to build and operate megaprojects involving cutting-edge technology or to take advantage of post-production learning opportunities.

In view of the importance of controlling greenhouse-gas emissions, it is appropriate that CTL plants receiving government incentives employ carbon-management approaches so that net greenhouse-gas emissions are at least comparable to those anticipated from the refinement and use of motor fuels derived through conventional production methods. CTL plants built in the vicinity of petroleum-bearing formations might be able to sell carbon dioxide for enhanced oil recovery operations. This would allow significantly lower federal subsidies, promote additional domestic production of petroleum, and provide an additional opportunity, beyond the Weyburn work in Canada, to examine the viability of large-scale, long-term carbon sequestration during enhanced oil recovery operations.

Another possible approach for carbon management from early CTL plants includes providing carbon dioxide to large-scale, long-term demonstrations of geologic sequestration, especially sequestration in deep saline formations. Finally, early CTL plants could serve as test beds and demonstrations of the cogasification of coal and biomass.

Demonstrate Carbon Sequestration

The main motivation for research and development directed at carbon sequestration in various types of geologic formations is the potential need to capture carbon dioxide emitted from central station power plants, especially those fired by coal. This purpose alone suggests that accelerated efforts are appropriate to establish the commercial viability, including the governing regulatory framework, of large-scale geologic sequestration of carbon dioxide.

As discussed in Chapter Three, geologic carbon sequestration is a critical technology for addressing the high level of plant-site carbon dioxide emissions associated with CTL production. Our analyses also indicate that the time required to bring carbon capture and sequestration to commercial viability will limit the maximum rate of CTL industrial development in the United States. For these reasons, we assign high priority to early and multiple large-scale, long-term demonstrations of geologic sequestration of carbon dioxide.

Our primary concern is not just technical viability, but commercial viability and public acceptance (see Chapter Six). Regarding technical viability, there is reason to be optimistic. According to the Intergovernmental Panel on Climate Change (2005),

> Observations from engineered and natural analogues as well as models suggest that the fraction retained in appropriately selected and managed geological reservoirs is very likely to exceed 99% over 100 years and is likely to exceed 99% over 1,000 years.

Moreover, no significant adverse findings have been reported in the moderate-scale carbon-sequestration field tests at In Salah, Algeria, the Sleipner gas field in the North Sea, and in the large-scale enhanced oil recovery test in the Weyburn oil field in Canada.

These positive findings aside, establishing the technology base required to safely and economically sequester billions of tons of carbon dioxide per year should be recognized as one of the great technical challenges of the next few decades. The U.S. Department of Energy Regional Carbon Sequestration Partnership Program is now transitioning to moderate- to large-volume injections over a three- to six-year period. Our findings suggest that priority be given to supplementing these tests with larger scale (multimillion tons per year) and longer duration (at least 30 years) demonstrations.

Develop Combined Coal and Biomass Gasification

Missing from the federal R&D portfolio are near-term efforts to establish the commercial viability of a few techniques for the combined use of coal and biomass. Such a combination offers significant cost and environmental payoffs, as discussed in Chapter Three. The most pressing need for near-term research centers on developing an integrated gasification system capable of handling both biomass and coal. The problem is to devise a system that grinds, pressurizes, and feeds a stream of biomass or a combination of biomass and coal into the gasifier with high reliability and efficiency and without damaging the gasifier. This is a fairly minor technical challenge—i.e., it is an engineering problem focusing on performance and reliability, not a science problem. To establish the design basis for such a system requires materials testing and the design, construction, and operation of one or a few test rigs. These test rigs need to be fairly large so that they are handling flows close to those that would be processed in a commercial plant. This is because solids are involved, and it is very difficult to predict performance and reliability of systems for handling and processing solids when the size or design throughput of the system undergoes a large increase.

Support High-Payoff R&D

At present, the federal government is supporting research on coal gasification and associated synthesis-gas cleaning and treatment processes. Most of this federally funded research is directed at nearer-term, lower-risk concepts for advanced power generation and the production of hydrogen, but much of it is also directly applicable to CTL production using either the FT or MTG approach.

If the federal government is prepared to promote early CTL production experience, expanded federal R&D efforts should be considered, especially R&D directed at high-risk, high-payoff opportunities for cost reduction and improved efficiency and environmental performance. Especially fruitful areas for longer-term R&D are oxygen production at reduced energy consumption, improved gas-gas separation technology, higher temperature gas-purification systems, and reduced or eliminated oxygen demand during gasification. As a collateral benefit of this public investment, longer-term research efforts would also support the training of specialized scientific and engineering talent required for future progress.

During the late 1970s, the entire federal R&D program in coal liquefaction was dedicated to advancing and demonstrating direct liquefaction. Approaches involving indirect liquefaction, such as the FT and MTG methods, received negligible, if any, support. Today, the situation is completely reversed. We are not aware of any federal support for research on direct coal liquefaction. Although the development risks of direct coal liquefaction are high, the direct approach offers the potential for considerable cost reductions, higher conversion efficiencies, and therefore less coal consumption per barrel of product and lower plant-site greenhouse-gas emissions.

Because the federal energy-technology portfolio is managed by resource categories (i.e., fossil energy, renewables, nuclear), there are few, if any, R&D efforts that capitalize on the opportunities associated with combined resources. We have already discussed one of them—the combined use of coal and biomass in an FT or MTG liquefaction process—but it is highly likely that additional opportunities are available. Productive research in this area could include the use of nuclear energy as a source of thermal energy or hydrogen in liquefaction, advanced approaches for very small–scale facilities for liquid-fuel production, and the integration of indirect coal liquefaction and biomass fermentation approaches to liquids production.

Overall, we suggest that the federal government examine the depth, breadth, and adequacy of resources devoted to its research portfolio for the production of fuels for transportation applications. The current R&D focus on alcohol fuels and hydrogen is too narrow, as is the emphasis on nearer-term rather than on longer-term opportunities.

Air Force Options for Coal-to-Liquids Industrial Development

Should the U.S. Air Force choose to play an active role in promoting the development of a domestic CTL industry, it should do so recognizing that the primary potential benefits of success would accrue more to the nation as a whole than to the U.S. Air Force as an institution. In particular, our analysis shows that the nation is justified in spending $6 to $24 per barrel more than the world oil price to bring alternative fuels online. However, the entire DoD uses only about 1.5 percent of the oil used by the United States. Applying our analysis only to DoD, we conclude that DoD, acting strictly in its own interests, is justified in spending for alternative fuels no more than $0.50 per barrel above world oil prices.[6]

The U.S. Air Force's 2016 goal of being prepared to purchase JP-8 blends containing unconventional fuels is consistent with an overall federal insurance-policy strategy. Specifically, if the potential fuel purchases associated with the U.S. Air Force goal were to be fully met with a 50/50 FT-JP-8 blend, the 50,000–80,000 bpd of required FT CTL capacity falls within the overall production requirements of an insurance-policy strategy—namely, obtaining early production experience from a limited number of CTL plants.

---

[6] From Chapter Five, the national benefit is based on the marginal net surplus. The DoD benefit is based on the marginal consumer surplus downsized to reflect the DoD fraction of total domestic liquid-fuel use.

Air Force Incentives for Early Industrial Experience

Our analysis of incentives to encourage early private investment in CTL production revealed that a balanced and cost-effective approach, from both government and private-sector perspectives, would include a price floor, an investment incentive (such as a tax credit or loan guarantee), and an income-sharing agreement. Approaches for implementing such a package could involve the U.S. Air Force and the Defense Logistics Agency. For instance, DoD could grant long-term contracts for military fuels or blend stocks. Such contracts could include price floors to protect private investors against low world oil prices and price discounts to allow income sharing during periods of high oil prices.

To be more cost-effective, fuel contracts designed to promote early CTL experience should be part of a broader package of investment incentives, such as investment-tax credits, accelerated depreciation, and loan guarantees. Alternatively, DoD could be a partial investor in new plants as well as a fuel purchaser. By reducing up-front costs to early investors, such direct DoD investments could serve the same role as the investment incentives mentioned. These additional instruments could allow lower price floors and lessen the probability of out-year government purchases at above-market prices.

Another option for the U.S. Air Force and DoD is to use DoD's contracting authority to establish a guaranteed or fixed price over a significant portion of the operating life of a CTL plant. Such agreements are rarely observed in contracts between private parties.[7] Our findings indicate that a long-term price guarantee is far less likely to serve federal government interests than are the alternative approaches described earlier.[8]

Policy Issues

   Contractual Limitations. Currently, DoD contracts are limited by law (10 USC 2306b) to a duration of no more than five years[9] and a total amount of less than $500 million, unless specifically authorized otherwise by Congress. As such, DoD's ability to provide incentives for private investments in early CTL plants is severely limited. Private investors would likely evaluate the viability of a CTL project using an operating life of at least 15 years; 30 years is not an uncommon planning factor. Only five years of protection against low world oil prices may not be sufficient to promote investment in CTL plants. The $500 million ceiling limits contracting authority to

---

[7]   When transaction prices under an agreement depart from market prices for long periods of time—say, several years—the parties to an agreement no longer share a mutual interest in the continuation of a fixed-price agreement. The longer such dissonance persists, the harder it becomes to prevent termination or renegotiation of the contract. As a result, long-term contracts with fixed prices rarely survive for the long term, particularly in markets for products with prices as volatile as those for petroleum.

[8]   Camm, Bartis, and Bushman, 2008, compared these alternatives directly and explained in more detail why a price guarantee is undesirable.

[9]   The law also allows five one-year options for a total of ten years.

fairly small amounts of coal-derived military fuels. For example, if the anticipated five-year average price of CTL fuels is $70 per barrel, the procurement-cost ceiling would limit procurement to less than 4,000 bpd. But such a procurement limit would provide a fairly weak incentive to pioneer commercial CTL plants that attempt to capture economies of scale by operating at much higher liquid-production rates. New legislative authority is needed if DoD and the U.S. Air Force are to overcome the limitations imposed on contract duration and size.

Technology Bias. Presently, two CTL technologies are ready for commercial applications: FT and MTG. The nation and DoD should be neutral with regard to how private-sector investments are apportioned between these two choices. Most large petroleum refineries can meet varying seasonal demand for gasoline versus middle distillates (such as diesel and jet fuel) by significantly shifting their product mixes. If a large domestic CTL industry develops and produces predominantly gasoline or predominantly middle distillates, these petroleum refineries would respond by appropriately shifting their product mixes to match domestic demand. For the nation and DoD specifically, the net benefits are the same—reduced demand for petroleum-derived fuels, lower world oil prices, and increased resiliency in the supply chain for liquid fuels.

To avoid technology bias by the government, federal efforts to promote early CTL commercial experience should include incentives beyond the procurement of military fuels.

Protection from Fuel-Supply Disruptions. The development of a commercial CTL industry in the United States would likely improve the resiliency of the petroleum supply chain (see Chapter Five) and thereby somewhat reduce the adverse consequences of fuel-supply disruptions caused by natural disasters, such as hurricanes. In the event of a sudden, large loss of oil supplies from one or more international sources, the U.S. Air Force would likely encounter very high fuel prices and considerable pressure, via Executive Orders, to minimize fuel use.

The extensive measures in energy efficiency and conservation that have been and continue to be implemented by DoD and the services serve to reduce the amount of fuel required for combat operations or to maintain readiness and thereby the impact of the sudden and large fuel-cost increases that would be encountered by a disruption to the world petroleum market. In the event that DoD is unable to receive needed fuels, the Defense Production Act (P.L. 81-774) contains provisions for performance on a priority basis of contracts for the production, refining, and delivery of petroleum products to DoD and its contractors (Swink, 2003).

If DoD and the U.S. Air Force believe that further measures may be needed to reduce vulnerability to fuel-supply disruptions, we suggest consideration of the costs and benefits of creating a DoD-operated military fuel reserve. Another approach is to establish agreements with traditional fuel suppliers that would grant priority to the U.S. Air Force, or DoD, during disruptions (Dalton, 2007). In fact, the U.S. Air Force has successfully used precisely such agreements to ensure priority access to air transport during national security contingencies.[10]

Addressing Budget Impacts of Fuel-Price Volatility. The U.S. Air Force and Navy interest in unconventional fuels is motivated in part by recent budgetary shortages created by unexpected, rapid increases in the price of fuels. A 2006 Office of the Secretary of Defense briefing on energy costs noted that the "recent dramatic increase in fuel costs was not programmed and puts pressure on [the] DoD budget. Further increases are likely. DoD is considering alternatives and options to alleviate pressure on [the] budget and increase energy security" (Illar, 2006). In a recent 12-month period, the U.S. Air Force saw its per-gallon price for JP-8 rise 31 percent, from $1.74 to $2.53 per gallon. The Air Combat Command found itself facing a shortfall of $825 million in must-pay funds (Wicke, 2006). In 2006, the U.S. Air Force paid more than $5.8 billion for jet fuel, more than twice the $2.6 billion it spent in 2003, even though 2006 fuel consumption was significantly below 2003 levels.

Supplemental funding has prevented such rising fuel costs from interfering with combat missions. But these dramatic increases have forced DoD to make adjustments elsewhere, e.g., in cutting flying hours in pilot-training programs, trimming personnel, and postponing repairs at military installations (Montgomery, 2007). Such changes can lead to lower combat readiness of aircrews and reduced quality of life for airmen (Wicke, 2006).

Although our research did not directly address instability of prices, we are able to make some observations about how to mitigate this problem. In particular, by using contracts that incorporate longer-term future prices, for either conventional or CTL-based fuels, whichever is less expensive, the U.S. Air Force could lock in prices during the programming phase of the Planning, Programming, Budgeting, and Execution System process. It could then employ those prices to buy conventional or unconventional fuels during the execution phase. Alternatively, DoD could adjust the revolving fund it has used for many years to manage changes in energy prices between a programming phase and an execution phase.[11] By recapitalizing the fund and changing operating targets in it, DoD could allow it to absorb larger unexpected fuel-price increases than it has apparently been able to accommodate in recent years. DoD could then use monies from the fund to buy conventional or CTL-based fuels, whichever would be more cost-effective.

---

[10] The Civil Reserve Air Fleet was the result of such an agreement. For details, see Chenoweth (1990).

[11] This fund has traditionally allowed components of DoD to pay the programmed fuel prices during a period of execution and then rebalance the fund through direct appropriations when any difference in prices occurred. If prices were higher than anticipated, direct appropriations could replenish the fund, which the components had drawn down. If prices were lower than anticipated, direct appropriations for follow-on budgets would be lower until the balance in the fund had dropped to a target level.

Scoping Federal Efforts: How Much Is Enough?

Implementing an insurance-policy strategy for CTL commercial development will involve government expenditures. These expenditures need to be formulated to be commensurate with expected benefits. Since the principal objective of the insurance-policy strategy is promoting early industrial experience, the primary benefit is acceleration of commercial development over what would be the case in the absence of government action. If we assume that the insurance-policy strategy promotes commercial development by five years (as compared to no federal program), the potential net benefit of that strategy would be roughly $50 billion to $150 billion, per the analysis presented at the conclusion of Chapter Five, with the lower end of this estimate associated with crude oil prices averaging $60 per barrel over the next 40 years and the upper end associated with crude oil prices averaging $100 per barrel during those years.

Three components of the insurance-policy strategy provide the foundation for the very existence of a CTL industry and therefore should be measured against the full value of domestic CTL industrial development—namely, hundreds of billions of dollars. These three components are large-scale, long-term demonstrations of carbon sequestration in the United States; research on dual-fired CBTL systems; and the development of scientific and engineering talent as a collateral consequence of public investment in longer-term, higher-risk R&D.[12]

Our analysis also provides a framework for scoping the funding of technology-development efforts, such as those associated with CBTL development and those that would be included in a high-risk, high-payoff R&D portfolio as in the insurance-policy strategy. If the outcome of successful technology development is lower-cost CTL production, the benefit is increased economic profits. Suppose that success in a particular technology-development effort or program leads to a $5 per barrel reduction in production costs for all CTL production occurring after 2023. With the same assumptions used in Chapter Five to estimate the value of a domestic CTL industry, this R&D-induced increment to economic profits has a net present value of roughly $25 billion. Given the risks associated with research, but also the collateral benefits, this figure argues for a multiyear CTL technology-development program with a net present value in the range of a few billion dollars.

---

[12] Public benefits from public investments in each of these three components of the insurance-policy strategy extend well beyond those associated with CTL development. For carbon-sequestration research and demonstration, much larger benefits derive from keeping open the option of coal use for electric-power generation. And scientists and engineers trained through public investment in research provide a foundation for innovation across all sectors of our economy.

A Stable Framework for Reducing World Oil Prices

Our analysis shows that important national benefits accrue from measures that lower worldwide demand for crude oil from the Persian Gulf and thereby reduce world oil prices. On a barrel-to-barrel basis, conservation reduces dependence on Persian Gulf crude oils and lowers world oil prices as effectively as enhanced supply does. A barrel not consumed has the same value as an extra barrel of domestic production. Just as there are many technological approaches for unconventional-fuel production, there is great variety in the opportunities for energy conservation. An effective and economically efficient energy policy is one that employs the power of the marketplace to choose among all competing alternatives.

Yet federal energy legislation and policies over the past 25 years have taken just the opposite approach. Instead of goal-directed, broad-based measures, U.S. energy policy has consisted of piecemeal measures, with legislation focusing on sector- and technology-specific subsidies, tax breaks, and regulations. The consequences are special interests before national interest and a history of ineffective and costly energy policies.[13]

As the federal government considers market-based measures for reducing greenhouse-gas emissions, we strongly suggest including consideration of market-based measures to reduce U.S. demand for crude oil and thereby to decrease world oil prices. Our analysis suggests that a tax on conventional petroleum of at least $10 per barrel (about $0.25 per gallon) would yield net national economic benefits. An oil tax, unlike government subsidies, tax breaks, or regulations, would not adversely discriminate among alternative conservation and unconventional-fuel opportunities. A modest oil tax could also help limit potential increases in petroleum dependence, as energy users respond to market signals to reduce greenhouse-gas emissions.

A $10 per barrel tax on domestic production and imports would yield government revenues of $75 billion per year. A portion of this revenue could be used to compensate families who would disproportionately bear the burden of an oil tax.

---

[13] Moreover, a sizable portion of these costs are incurred by lower- and middle-income families. A recent report by the Brookings Institution observed that "command-and-control systems [i.e., regulatory as opposed to tax approaches to reduce energy use] can have adverse distributional consequences that are both hidden and difficult to remedy" (Furman et al., 2007).

Cost-Estimation Methodology and Assumptions

This appendix summarizes the methodology and assumptions that underlie the estimated costs of producing naphtha and diesel from a large first-of-a-kind FT CTL
plant, as discussed in Chapters Three and Seven. Production costs are calculated using
a discounted-cash-flow model in which both capital and operating costs are entered
in first-quarter 2007 dollars. All references to crude oil prices are based on West Texas
Intermediate as a benchmark price, stated in January 2007 dollars. The calculated production costs presented in Chapter Three are annualized costs per barrel, adjusted to
reflect product premiums over crude oil, and are expressed as crude oil equivalent.

Capital and Operating Cost Estimates

For our calculations, capital costs are defined as all outlays made after the decision to
construct a commercial plant and prior to the production of salable products. Capital
costs, as defined in this approach, do not include factors to account for either inflation or interest accrued during construction. The effects of interest payments during
construction are accounted for in the discounted-cash-flow calculations. For an FT
CTL plant, our capital costs do not include investments associated with developing the
mine or mines that will supply coal to the plant. Instead, we assume that coal will be
purchased via long-term contracts at prices that account for the costs associated with
developing and operating a coal mine.

Capital costs include two basic categories: plant costs and start-up costs.[1] Plant
costs dominate in our calculations and include all site preparation; design and construction costs for the entire plant, including on-site product upgrading required to
produce a finished diesel fuel and a naphtha product suitable for pipeline transport
to a refinery; auxiliary systems required for pollution control and coal-ash disposal;
and infrastructure required to obtain access to electric power, water, and coal and to
transport the products produced by the plant. For CTL plants built in the West, plant

---

[1] Land-acquisition costs are negligible when compared with plant costs and are ignored in our analysis.

119

costs could include substantial additional investments in infrastructure, such as roads, housing of construction workers, and water access.

In our calculations, all start-up costs, including home office costs, are expensed in the first operating year. Since these costs constitute a small fraction of total costs, a more detailed tax treatment of these items does not significantly influence our estimates.

Net plant-operating costs are the costs associated with operating and maintaining the plant minus any income produced from the sale of by-products, such as sulfur and ammonia, or coproducts, such as electricity and possibly carbon dioxide. We assume a CTL plant that is not configured for production of marketable ammonia or associated products. Sulfur sales do not significantly affect fuel-production costs and accordingly are ignored in our analysis. FT CTL plants do produce significant amounts of salable electricity. We assume that this electric power will displace baseload power from a new coal-fired power plant. The estimate presented in this appendix assumes that between 85 and 90 percent of the carbon dioxide that would be released into the atmosphere at the plant site is instead captured, compressed to 2,000 psi, and available at the plant boundary for pipeline transport. Other costs associated with carbon dioxide transport beyond the plant and sequestration are not included. A companion volume (Camm, Bartis, and Bushman, 2008) includes analysis of alternative assumptions regarding additional costs for the ultimate disposal or additional income from the sale of this captured and compressed carbon dioxide.

Key Financial Assumptions

Table A.1 summarizes the key parameters in the cost estimates for a CTL plant as presented in Chapter Three and used in Chapter Seven. The reference-case cost estimate in Chapter Three of $55 per barrel as the required crude oil selling price is based on the low end of the capital and operating cost ranges shown in Table A.1. Likewise, the high-plant-cost case estimate of $65 per barrel as the required crude oil price is based on the upper end of the capital and operating cost ranges.

To relate prices for refined products to those for crude oil, we follow Gray (SSEB, 2006, Appendix D) and assume that one barrel of FT zero-sulfur, high-cetane diesel is priced at a factor of 1.3 above the value of one barrel of West Texas Intermediate. We also assume that one barrel of FT naphtha would be priced at 93 percent per barrel of West Texas Intermediate, or, equivalently, 71 percent of FT diesel. We calculate the total liquids production in terms of DVE by multiplying naphtha production by 0.71 and adding the result to FT diesel production.

For the purposes of developing a product cost estimate, the operating life of the plant is set at 30 years. Estimates of product costs are highly insensitive to further increasing the financial time frame. Total plant costs are depreciated over a seven-year

Table A.1
Product Price-Calculation Assumptions

Assumption	Value
Capital investment (millions 2005 dollars)	3,300–4,050
Total depreciable plant costs (millions 2005 dollars)	2,960–3,700
Expenditure schedule for total plant costs (%)	
Year 1	5
Year 2	15
Year 3	32
Year 4	28
Year 5	20
Initial operating year (year 6)	6
Plant financial operating life (years)	30
Coal use (tons per barrel DVE)	0.55
Coal costs (dollars per ton)	30
Other variable costs (dollars per barrel DVE)	2.69
Annual fixed operating costs (millions of dollars)	132–176
Depreciation schedule for total plant costs	MACRS (200% declining balance)
Federal corporate tax rate (%)	34
State corporate tax rate (%)	3.2
Rate of return (real, after tax) (%)	10
Plant utilization rate (online factor) (%)	
Initial operating year	70
Years 2 through 30	90
Plant outputs	
FT diesel (bpd)	24,359
FT naphtha (bpd)	11,398
Total liquids (DVE bpd)	32,502
Electricity (MW)	204
Product prices: reference case, West Texas Intermediate	
Diesel-to-crude price ratio (barrel/barrel)	1.3
Naphtha-to-crude price ratio (barrel/barrel)	0.92
Electricity sale price (dollars per kWh)	0.05

recovery period, utilizing the double–declining balance/straight-line method allowed under the Modified Accelerated Cost Recovery System (MACRS).

Estimates of production costs provided are highly sensitive to the after-tax real rate of return and the plant online factor. For the per-barrel cost estimates in Chapter Three, we use a real after-tax rate of return of 10 percent. We believe that this rate of return is appropriate for large technology firms, such as major oil or chemical companies, that view investment in a first-of-a-kind CTL plant as an opportunity to establish a market position and a basis for improving the costs and performance of future plants. We note that this is not a rate of return that will attract investors that are not taking this longer view. To assess the effects of uncertainty about future oil prices, Chapter Seven takes a different perspective that allows consideration of a broader range of rates of return.

The plant online factors of 70 percent in the first operating year and 90 percent thereafter assume extensive and highly competent oversight of plant design, including evaluation and selection of processes, components, and materials.

Greenhouse-Gas Emissions: Supporting Analysis

This appendix documents our analysis of the greenhouse-gas emissions that we anticipate from commercial FT CTL and FT CBTL facilities and compares these estimates to greenhouse-gas emissions from conventional petroleum–based fuels. We take a total-fuel-cycle approach and account for the primary greenhouse-gas emissions that occur in the production and delivery of a fuel to the tank of a vehicle and in the combustion of that fuel in the vehicle's engine. Our analysis is based on preliminary designs of CTL facilities (SSEB, 2006, Appendix D; NETL, 2007b) and the data and methodology established for the Greenhouse Gases, Regulated Emissions, and Energy Use in Transportation (GREET) model, developed at the Argonne National Laboratory (ANL) (Wang, 1999; ANL, 2007). The analysis covers carbon dioxide, methane, and nitrous oxide emissions. Our results are presented in terms of CDE using the 100-year global warming–potential estimates of the Intergovernmental Panel on Climate Change (2007c).

Both FT and conventional fuels are treated on a common basis. For FT CTL production, emissions of greenhouse gases due to coal mining consist of carbon dioxide and nitrous oxide emitted by mining equipment and methane emitted during coal-mining operations. We account for greenhouse-gas emissions that would be associated with production of a near-zero-sulfur diesel, a low-sulfur reformulated gasoline, and electric power. Specifically, we include an estimate of the greenhouse-gas emissions associated with manufacturing a reformulated gasoline from the highly paraffinic, low-octane naphtha that is a direct product of FT synthesis. The greenhouse-gas emissions for transportation of the FT fuels to a vehicle are those for comparable products in GREET. In calculating the emissions during combustion, we assume complete combustion of the FT fuels so that all of the carbon in the fuel is emitted as carbon dioxide.

To establish a baseline for comparison, we also calculate the greenhouse-gas emissions that would result from the production and use of a similar mix of diesel and gasoline produced from conventional petroleum. The parameters for these estimates are drawn from the default values of GREET. We use values from GREET rather than running a scenario within GREET because GREET performs many auxiliary calculations that would make a direct comparison difficult. We apply values for emissions of carbon dioxide, nitrous oxide, and methane from the recovery of light crude oil, its transport to U.S. refineries, its refining and upgrading into low-sulfur diesel and reformulated gasoline, and its transportation to and combustion in a vehicle. Table B.1 lists relevant properties of the conventional and FT fuels considered in this analysis.

We do not account for secondary emissions of greenhouse gases that would be included in a more detailed life-cycle assessment. For example, we do not consider the emissions associated with building the CTL plant, building the refinery, or manufacturing the mining equipment. Also excluded are refinery losses associated with the production of fuels used in mining or transport. These types of secondary emission sources account for a small marginal increase in emissions of greenhouse gases from FT and conventional fuels (Wang, 1999).

Greenhouse-Gas Emissions of Conventional Fuels

Greenhouse-gas emissions from petroleum-based diesel and reformulated gasoline are derived from the default values from GREET 1.8a (ANL, 2007), as described here and summarized in Table B.2.

Oil Recovery

Greenhouse-gas emissions from oil recovery are estimated by GREET at 11.6 pounds CDE per million Btu (lower heating value [LHV]) of oil recovered.[1] This estimate assumes conventional primary oil-recovery methods. It includes 6.6 pounds of carbon dioxide, with the remainder representing emissions of methane and nitrous oxide. This amount also includes methane releases from the recovery of associated gases. Assuming a refining efficiency of 87.0 percent (see Fuel Production later in this appendix) and converting to higher heating values (HHV), we calculate that crude oil production causes greenhouse-gas emissions of 12.5 pounds CDE per million Btu (HHV) of ULSD fuel. For reformulated gasoline, we assume a refinery efficiency of 85.5 percent. This results in oil recovery contributing 12.7 pounds CDE per million Btu (HHV) of reformulated gasoline.

Crude Oil Transportation

For transportation of crude oil to refineries, the default value from GREET is 1.8 pounds CDE per million Btu (LHV) of oil delivered. The default value assumes a mix of imports, Alaskan, and lower-48 crude oils. Of this amount, 0.1 pound is attributed to emissions of methane and nitrous oxide. Considering refining losses and converting to higher heating values, we estimate that crude oil transportation accounts for

---

[1] GREET calculations are based on the fuels' lower heating values. Our calculations use higher heating values, consistent with general practice. This distinction does not significantly change relative or absolute greenhouse-gas emission calculations for oil and coal-derived liquids.

Table B.1
Selected Properties of Conventional Fuels, Fischer-Tropsch Diesel, and Fischer-Tropsch Naphtha

Fuel	Higher Heating Value (Btu per gal.)	Lower Heating Value (Btu per gal.)	Density (pounds per gal.)	Carbon content (pounds per gal.)
ULSD	138,500	129,500	7.07	6.16
Reformulated gasoline	121,800	113,600	6.24	5.24
LPG	91,000	85,000	4.24	3.48
FT diesel	127,300	118,500	6.29	5.34
FT naphtha	119,600	111,500	5.93	5.01

SOURCES: Brinkman et al. (2005), White (2007), ANL (2007).

greenhouse-gas emissions of about two pounds CDE per million Btu (HHV) of ULSD or reformulated gasoline.

Fuel Production

For petroleum, the GREET default values pertain to refining a light crude oil. Nearly all greenhouse-gas emissions from refining are due to carbon dioxide releases during oil-processing operations. For ULSD fuel production, the GREET default values are based on a refinery efficiency of 87.0 percent (LHV) and yield a refining contribution of 28.0 pounds CDE per million Btu (LHV). For reformulated-gasoline production, the GREET default lower heating value refining efficiency is 85.5 percent, which results in greenhouse-gas emissions of 31.7 pounds CDE per million Btu (LHV). On a higher heating value basis, these estimates become 26.2 pounds CDE per million Btu of ULSD and 29.6 pounds CDE per million Btu of reformulated gasoline.

Product Transportation

Product transportation covers moving diesel and gasoline from refineries to distribution terminals to final outlets, such as retail service stations. The default values in GREET for greenhouse-gas emissions due to transportation yield 1.1 pounds CDE per million Btu (HHV) of ULSD and one pound CDE per million Btu (HHV) of reformulated gasoline.

Product Combustion

Assuming complete combustion of the fuels and using Table B.1, we calculate 163.0 and 157.7 pounds of CDE emissions per million Btu (HHV) of, respectively, ULSD and reformulated gasoline.

Fischer-Tropsch CTL Fuels

For CTL facilities, we consider cases 3 and 8 of the plant designs prepared by Gray and associates at Noblis (SSEB, 2006, Appendix D) for the Southern States Energy Board. Both of these cases cover moderate-size facilities producing at a nominal rate of 30,000 bpd. Case 3 accepts bituminous coal as a feedstock and produces 187,500 million Btu per day of liquid fuels, and case 8 accepts subbituminous coal as a feedstock and produces 184,600 million Btu of liquid fuels. These two plant designs include capture and compression of slightly more than 85 percent of on-site carbon dioxide emissions. According to Gray and White (2007), the electricity required to compress the captured carbon dioxide in these cases is 53 MW.

In this appendix, we examine two other options for greenhouse-gas management. The first option assumes no control of greenhouse gases. Since carbon dioxide is released into the atmosphere, we adjust the preceding two design cases so that they do not capture and compress carbon dioxide. The primary effect of this adjustment is that an additional 53 MW of electric power are available for sale on the grid. We also assume that all of the electricity sold by the CTL plants is displacing electric power and carbon dioxide emissions that would otherwise be delivered from new baseload coal-fired power plants that do not control carbon dioxide emissions. In the second option, we assume that market-based measures to control greenhouse-gas emissions are in effect and that these measures are sufficient to motivate the capture and sequestration of carbon dioxide emissions from new coal-fired power plants. As discussed in Chapter Three, such measures should be more than sufficient to motivate carbon capture and sequestration at FT or MTG CTL facilities. For the analysis in this second option, we assume that 90 percent of plant-site emissions are captured and sequestered. For electricity sales, we examine two options: no carbon credit for electricity sales or a carbon credit. Providing no carbon credit for electricity sales is consistent with the assumption that the electricity coproduced by a CTL plant would displace power that would be generated by a new coal-fired plant that would also employ carbon capture and sequestration. Over the next few decades, electricity coproduced by a CTL plant may actually displace power produced by existing coal-fired power plants, justifying consideration of providing a carbon credit for electricity sales.

To allow comparison with the conventional-petroleum case, we account for the carbon dioxide emissions associated with upgrading the FT naphtha to a gasoline blendstock suitable for manufacturing reformulated low-sulfur gasoline. Table B.2 summarizes our results for fuels derived from conventional petroleum and from a CTL plant, including upgrading. The following sections describe our analytic approach.

Naphtha Upgrading

To obtain an estimate of the carbon dioxide emissions resulting from naphtha upgrading, we apply the results of Gregor and Fullerton (1989), who presented a design for

Table B.2
Fuel-Cycle Greenhouse-Gas Emissions of Conventional and Fischer-Tropsch Liquid Fuels
(pounds CDE per million Btu [HHV] of liquid fuels)

Life-Cycle Step	ULSD	Reformulated Gasoline	Bituminous Coal		Subbituminous Coal	
			No CCS	CCS	No CCS	CCS
Extraction and mining	12.5	12.7	37.8	37.8	9.4	9.4
Feedstock transportation	2.0	2.0	0.0	0.0	0.0	0.0
Refining and production	26.2	29.6	309.7	31.0	303.1	30.3
Credit for exported electricity	0	0	−55.5	−57.5 to 0	−43.5	−40.8 to 0
Product transportation	1.1	1.0	1.0	1.0	1.0	1.0
Product combustion	163.0	157.7	156.3	156.3	156.3	156.3
Total	204.7	203.0	449.4	168.7 to 226.2	426.3	156.3 to 197.0
Ratio to petroleum-derived fuels			2.2	0.83 to 1.1	2.1	0.77 to 0.96

NOTE: CCS = carbon capture and sequestration. Columns may not add due to rounding.

a stand-alone FT naphtha–upgrading plant that produces gasoline with a research octane number of 100.[2] This upgrading facility accepts 40,000 bpd of FT naphtha and produces 31,600 bpd of gasoline. The upgrading plant also produces 636 tons per day of a fuel gas that is similar in composition to LPG. The plant consumes 201 tons of this fuel gas and 230 megawatt-hours (MWh) per day of electricity. Carbon dioxide emissions from consumption of fuel gas within the upgrading plant and from the generation of required electricity constitute the greenhouse-gas emissions associated with upgrading FT naphtha.

To estimate the emissions of carbon dioxide due to FT naphtha upgrading, we scale the inputs and outputs of the 40,000 bpd upgrading plant to the FT naphtha levels produced by the Noblis case 3 and 8 plants, as shown in Table B.3.[3] The carbon

---

[2] The research octane number refers to a specific test procedure. The octane rating reported at a retail pump would generally be at least five points lower.

[3] This scaling is based on either the assumption that transport of FT naphtha to a large centralized upgrading facility would result in negligible greenhouse-gas emissions or the assumption that scaling down an FT naphtha–upgrading facility would not result in significant energy losses.

dioxide emissions shown in Table B.3 result from the combustion of the fuel gas consumed during upgrading.

To calculate an overall carbon balance for both CTL production and FT naphtha upgrading, we assume that surplus electricity from the CTL plants is used to provide the required electric power, which reduces the net amount of electricity available for export (and consequently, the carbon credit from exported electricity in the non–carbon capture and sequestration cases), as shown in Table B.4. Also shown in Table B.4 is the calculated energy content of the final slate of liquid products—namely, FT diesel, gasoline, and a small amount fuel gas that we assume would be marketed as LPG.

Fuel-Cycle Greenhouse-Gas Emissions

The methods for calculating the emissions of greenhouse gases associated with each step of the process of mining, transporting, producing, upgrading, and delivering the FT liquids, as shown in Table B.2, are described next.

Coal Mining. More than 60 percent of U.S. bituminous-coal production is based on underground mining. Assuming that the bituminous coal used in the Noblis case 3 analysis is produced in an underground mine, we estimate that greenhouse-gas emissions would be 16.2 pounds CDE per million Btu (HHV) of coal mined. Of this amount, 1.9 pounds are from emissions associated with operating mining equipment, and 14.3 pounds are from methane released during mining operations. The estimate for mining equipment (i.e., nonmethane) emissions is based on GREET default values. The methane-emission estimate is based on the 2006 *Emissions of Greenhouse Gases*

Table B.3
Estimated Performance and Emissions for Naphtha Upgrading

Plant	Reference Plant	Bituminous CTL (Case 3)	Subbituminous CTL (Case 8)
Naphtha input (bpd)	40,000	11,400	11,220
Gasoline out (bpd)	31,600	9,000	8,860
Fuel gas out (tons per day)	435	124	122
Fuel gas consumed (tons per day)	201	57.3	56.4
Carbon dioxide emissions (tons per day)[a]	605	172	170
Electricity consumed (MWh per day)	234	66.8	65.7
Liquid-energy loss[b] (million Btu per day)		5,850	5,760

[a] Calculated by assuming that the carbon content of the fuel gas is 82 percent, consistent with the value listed in Table B.1.

[b] Liquid-energy loss is the difference between the energy of the naphtha entering the upgrading plant and the gasoline and LPG leaving the plant.

Table B.4
Net Products for Fischer-Tropsch Coal-to-Liquids Plus Naphtha Upgrading

Coal and CCS Option	Bituminous Coal		Subbituminous Coal	
	No CCS	CCS	No CCS	CCS
CTL plant power exports (MWh per day)[a]	6,168	4,870	4,776	3,421
Net power exports (MWh per day)	6,101	4,803	4,710	3,355
Energy content of CTL liquids (million Btu HHV per day)	187,500	187,500	184,600	184,600
Energy content of final liquids (million Btu HHV per day)	181,600	181,600	178,800	178,800
Carbon content of final liquids (tons carbon per day)	3,871	3,871	3,811	3,811

NOTE: CCS = carbon capture and sequestration.

[a] The CTL plant electric-power exports for the cases with carbon capture and sequestration have been adjusted to account for the additional amount of carbon dioxide requiring compression to achieve 90-percent capture, as compared to Noblis cases 3 and 8 (SSEB, 2006, Appendix D).

Report (EIA, 2007e).[4] From the Noblis case 3 analysis and Table B.4, each million Btu (HHV) of liquid product requires that 2.34 million Btu (HHV) of bituminous coal be received at the CTL plant, resulting in mining-related greenhouse-gas emissions of 37.8 pounds per million Btu (HHV) of FT-liquid fuels.

Nearly all subbituminous coals are produced using surface mining methods. For subbituminous CTL plants, we estimate that greenhouse-gas emissions from coal production would be 4.3 pounds per million Btu (HHV) of coal mined. Of this, we estimate that 1.8 pounds would be due to nonmethane emissions associated with mining equipment, per the GREET default values, and 2.5 pounds are from methane released during mining operations.[5] From the Noblis case 8 analysis and Table B.4, each mil-

---

[4] EIA (2007e, Table 15) reported that the mining of 358 million tons of coal produced from underground mines in 2006 caused the release of 50.5 million tonnes CDE of methane for 2006. This estimate is based on a global-warming potential for methane of 23. Correcting for the revised global-warming potential of methane of 25 (IPCC, 2007c), converting to English units, and dividing by 2006 underground coal production, we estimate methane emissions of 338 pounds CDE per ton of underground-mined bituminous coal. The higher heating value of the bituminous coal as received at the CTL plant is based on Gray's case 3: 11,800 Btu per pound.

[5] EIA (2007e, Table 15) reported that the mining of 803 million tons of coal produced from surface mines in 2006 caused the release of 14.2 million tonnes CDE of methane for 2006. Correcting for the revised global-warming potential of methane (IPCC, 2007c), converting to English units, and dividing by 2006 surface-mined coal production, we estimate methane emissions of 42.4 pounds CDE per ton of surface-mined subbituminous

lion Btu (HHV) of liquid product requires that 2.19 million Btu (HHV) of subbituminous coal be received at the CTL plant, resulting in mining-related greenhouse-gas emissions of 9.4 pounds per million Btu (HHV) of FT-liquid fuels.

Coal Transportation

We assume that the CTL facility is built at or sufficiently close to one or more mines such that greenhouse-gas emissions due to coal transportation from the mine to a CTL plant are negligible, as compared to CTL production and upgrading.

Coal-to-Liquids Production and Upgrading

The bituminous CTL plant consumes 17,987 tons of coal per day representing an energy input of 424,493 million Btu (HHV) per day and a carbon input of 11,541 tons per day (SSEB, 2006, Appendix D). From Table B.4, the carbon content of the final liquid products is 3,871 tons per day, with an energy content of 181,600 million Btu. From these parameters, we estimate that the net emissions for the bituminous CTL plant (including naphtha upgrading) without carbon capture and seqestration would be 309.7 pounds CDE per million Btu (HHV) of liquid fuel produced. Assuming 90-percent capture and sequestration of plant site emissions (including those associated with naphtha upgrading), we estimate that the bituminous coal plant with carbon capture and sequestration would release 31.0 pounds CDE per million Btu (HHV) of liquid fuel produced.

For the subbituminous CTL plant analyzed by Gray, 22,988 tons of coal per day are consumed, representing an energy input of 390,796 million Btu (HHV) per day and a carbon input of 11,200 tons per day (SSEB, 2006). Using these parameters and Table B.4, we estimate that the net emissions from a subbituminous CTL plant (including naphtha upgrading) without carbon capture and seqestration would be 303.1 pounds CDE per million Btu (HHV) of liquid fuel produced. Assuming 90-percent capture and sequestration of plant-site emissions (including those associated with naphtha upgrading), we estimate that the subbituminous CTL plant with carbon capture and sequestration would release 30.3 pounds CDE per million Btu (HHV) of liquid fuel produced.

Credit for Exported Power

To calculate displaced carbon dioxide emissions, we assume a carbon dioxide–emission factor of 2.09 pounds per 10,000 Btu of coal consumed, based on EIA data for 2006 (EIA, 2008d, Tables 2.1f, 12.3). For CTL plants not employing carbon capture and sequestration, we assume that the coproduced electricity will displace power from a new coal-fired power plant operating at an average annual heat rate of 7,900 Btu per

---

coal. The higher heating value of the subbituminous coal as received at the CTL plant is based on the Noblis case 8: 8,500 Btu per pound.

kWh, which is possible from an advanced IGCC plant with a design heat rate of 7,600 Btu per kWh.[6] (See case 3D in Buchanan, Schoff, and White, 2002). These assumptions yield a carbon dioxide credit of 1.65 pounds per kWh of electricity sold by the CTL plant.

For CTL plants employing carbon capture and sequestration and displacing power from existing coal-fired power plants, we assume that the displaced power would have been produced at a heat rate of 10,400 Btu per kWh, which is the 2006 average for all coal-fired power plants operating in the United States (EIA, 2008d, Tables 2.1F, 8.2d). For this option, the carbon dioxide credit is 2.17 pounds per kWh of electricity sold by the CTL plant.

The credit per million Btu of final liquid products, after naphtha upgrading, is derived using the Table B.4 values for net power exports and energy contents of the final liquid products. For the two cases with carbon capture and sequestration, Table B.2 includes the options of allowing or disallowing a credit for exported power. This results in a range of values for both total life-cycle emissions and the comparative ratios.

Product Transportation

We assume the same emission rate for the transportation of the upgraded FT liquids as for low-sulfur diesel and petroleum gasoline. We do not allocate a transportation charge for the LPG coproduct. LPG represents a small fraction of total production, and omitting transportation-related emissions will not affect our calculations.

Product Combustion

We assume complete combustion of the FT diesel, FT-derived gasoline, and LPG during use. Using Table B.4, we multiply the total carbon in the product by 3.67 to convert to mass of carbon dioxide emissions and divide by the total energy in terms of the higher heating values of the products.

Greenhouse-Gas Emission Summary and Comparative Ratios

The results of our analysis are presented in Table B.2. The calculations presented in Table B.2, as well as the supporting Tables B.3 and B.4, pertain to two cases of a single CTL design study (SSEB, 2006, Appendix D). More detailed design efforts could yield slightly different results. Likewise, greenhouse-gas emissions for fuels derived from conventional petroleum are highly dependent on the crude oil source, refining methods, and quantity and quality of other petroleum products being produced at the refinery. When comparing conventional petroleum–derived fuels with coal-derived liquid

---

[6] The advanced power plant is based on case 3D (IGCC plant) in Buchanan, Schoff, and White (2002), which has a design heat rate of 7,599 Btu/kWh. The current state of the art for IGCC power generation is a design heat rate of about 8,400 Btu/kWh (NETL, 2007c).

fuels, uncertainties on both sides of the ledger need to be acknowledged. These uncertainties suggest that no more than two significant figures should be used in comparative analyses, consistent with the presentation of the ratios in Table B.2.

Two important findings are revealed by our analysis. First, without any carbon management, the production and use of fuels produced in early CTL plants can be expected to cause slightly more than a doubling of greenhouse-gas emissions, as compared to production and use of the same amount of fuels derived from conventional crude oils. Second, with carbon capture and sequestration at the 90-percent level, the production and use of fuels produced in early CTL plants should not cause a significant increase in total-fuel-cycle greenhouse-gas emissions. Moreover, if electricity sold by a CTL plant is used to displace power at existing coal-fired power plants that are not capturing and sequestering carbon dioxide emissions, slight reductions—10 to 25 percent—in total-fuel-cycle greenhouse-gas emissions would likely result, as compared to production and use of the same amount of fuels derived from conventional crude oils.

Our analysis of the two CTL plant designs by Noblis indicate that the subbituminous CTL plant has, on a total-fuel-cycle basis, lower greenhouse-gas emissions than the bituminous CTL plant. This finding should not be generalized. Much of the difference between the two estimates stems from the release of methane during underground coal mining. While it is true that, on average, bituminous-coal production generates higher methane emissions than subbituminous coal production, approaches are available for capturing methane prior to mining. Additionally, the higher plant-site emissions for bituminous CTL plants (310 versus 303 pounds per million Btu of liquid fuels) appear to be a consequence of technology choices and the design philosophy employed by the Noblis team. In particular, we see no reason to anticipate that CTL plants based on bituminous coal will be less efficient or yield higher plant-site greenhouse-gas emissions, as compared to plants designed to accept subbituminous coals. More likely, just the opposite will be the case, especially if plants operating on subbituminous coals are required to minimize water use.

Carbon Balance for Fischer-Tropsch Coal/Biomass-to-Liquids Production

In this section, we calculate the carbon balance for FT liquids produced in a hypothetical facility processing a combination of coal and biomass and capturing and sequestering plant-site emissions of carbon dioxide, as discussed in Chapter Three. For our calculation, we assume that the energy contents of the biomass and coal fed to the facility are equal. Our calculations apply specifically to switchgrass, but our findings should be equally valid for other sources of biomass that do not require extensive energy expenditures during cultivation. In particular, our results should also apply to corn stover, other agricultural residues, and forest residues.

Farrell et al. (2006) accounted for emissions incurred while farming the biomass, including emissions associated with operating farm and irrigation equipment, production and transport of fertilizer and agrochemicals, and decomposition of a portion of the nitrogen in the fertilizer into nitrous oxide. They did not account for emissions of carbon dioxide due to indirect land-use change; this phenomenon may result when biomass production displaces food production, forcing the conversion to farmland of other lands, possibly in other countries, with stores of soil carbon. On conversion to farmland, this carbon that is stored in soil could be released as carbon dioxide and may be attributed to the biomass production that induces that conversion to farmland. Several recent studies have attempted to account for emissions due to indirect land-use change (Fargione et al., 2008; Searchinger et al., 2008) and concluded that, depending on the land that is converted to farmland, the emissions of carbon dioxide due to induced land-use change can dominate the life-cycle emissions of the fuel derived from the biomass feedstock. Nor did Farrell et al. account for the possibility of soil carbon storage, which results when certain types of grasses are grown on degraded lands (Tilman, Hill, and Lehman, 2006).

For the biomass feedstock, we apply data that Farrell et al. have made available (Farrell et al., 2006) for their cellulosic case. We assume a carbon content of 47 percent for dry switchgrass and a higher heating value of 8,000 Btu per pound, consistent with values reported in the literature (Energy Research Center of the Netherlands, undated; EERE, 2006). Relevant data used in the analysis appear in Table B.5.

To develop a rough estimate of greenhouse-gas emissions, we assume that a combined CBTL plant will have the same overall higher heating value energy efficiency as a small CTL plant, such as the Noblis case 1 (SSEB, 2006, Appendix D), which produces the energy equivalent of 10,790 bpd of diesel fuel from bituminous coal.

Table B.5
Properties of Switchgrass Used in Coal- and Biomass-to-Liquids Carbon Balance Calculation

Property	Value
Carbon content (dry basis) (%)	47
Higher heating value (Btu per pound)	8,000
CDE emissions during cultivation (pounds carbon dioxide per pound dry switchgrass)	0.07
CDE emissions during transportation (pounds carbon dioxide per pound dry switchgrass)	0.02
Total CDE emissions for production and delivery (pounds carbon dioxide per pound dry switchgrass)	0.09

SOURCE: Farrell et al. (2006).

To maintain a 50/50 coal-biomass energy split, the CBTL plant would accept 2,693 tons of coal per day and 3,972 tons of dry switchgrass per day. Daily carbon input would be 3,613 tons, of which 1,746 tons derive from coal and 1,867 derive from the switchgrass feed to the plant.

We assume that the CBTL plant (and an associated naphtha-upgrading plant) would produce the energy equivalent of 10,360 bpd of FT diesel fuel containing 1,157 tons of carbon[7] and emit (in the case with no carbon capture) 2,456 tons of carbon per day in the form of carbon dioxide. We assume that the plant, including the upgrading plant, captures and sequesters 90 percent of the carbon dioxide from all process streams, so that emissions are 900 tons of carbon dioxide per day. Relevant fuel properties used in the analysis are listed in Table B.1.

Summary of Coal/Biomass-to-Liquids Fuel-Cycle Greenhouse-Gas Emissions

Carbon flows and fuel-cycle greenhouse-gas emissions are summarized in Table B.6. All entries are CDE per gallon of fuel (FT diesel energy equivalent) produced at the CBTL plant and are calculated based on the following assumptions.

Coal Mining and Switchgrass Cultivation and Transportation. Coal is produced from an underground mine, resulting in greenhouse-gas emissions of 16.2 pounds CDE per million Btu (HHV) of mined coal. Greenhouse-gas emissions (CDE) due to coal mining and transportation are applied from the earlier discussion of CTL. CDE emissions due to cultivation and transportation of switchgrass are listed in Table B.5.

Biomass Carbon Credit. Through photosynthesis, the switchgrass converted 6,845 tons of carbon dioxide into 1,867 tons of carbon. The carbon in the switchgrass forms a closed loop with the atmosphere and reduces net emissions of carbon dioxide by 6,845 tons per day or, equivalently, 31.5 pounds per gallon of upgraded FT liquids.

Coal/Biomass-to-Liquids Production. The daily greenhouse-gas emissions of the plant are 246 tons of carbon, equivalent to 900 tons of carbon dioxide. Each day, 2,210 tons of carbon, equivalent to 8,103 tons of carbon dioxide, are sequestered.

Credit for Exported Power. Under the option that power sold by the CBTL plant will displace electricity generated by existing coal-fired power plants, the export credit would be 2.17 pounds CDE per kWh of exported power. The small CTL plant produces 27 MW of electric power for sale to the grid (SSEB, 2006, Appendix D, case 1). Assuming that electric-power requirements for preparing biomass will be met through reduced power for coal pulverization, the net cogenerated power of the CBTL plant should be similar to that of the reference CTL plant. The net credit for the CBTL plant is 703 tons CDE per day or, equivalently, 3.2 pounds per gallon of upgraded FT liquids.

---

[7]   For the hypothetical CBTL plant, the liquid yield and the carbon content of that yield are based on 96 percent of the values reported for the case 1 analysis (SSEB, 2006, Appendix D). This factor accounts for liquid losses during naphtha upgrading.

Table B.6
Estimated Carbon Balance for Fischer-Tropsch Coal- and Biomass-to-Liquids Plant with Carbon Capture and Sequestration

Process Step	Carbon Flows (pounds CDE per gallon FT diesel)		
	To CBTL Plant	To Atmosphere	Captured
Switchgrass cultivation and transportation		1.7	
Switchgrass carbon	31.5	−31.5	
Coal mining and transportation		2.4	
Coal carbon	29.4		
Captured and sequestered carbon			37.3
CBTL plant-site emissions		4.1	
Credit for exported electricity		−3.2 to 0	
Product delivery		0.1	
Fuel carbon released during combustion		19.5	
Net emissions		−6.8 to −3.6	

NOTE: Assumes equal energy input from coal and dry biomass.

Product Transportation. Carbon dioxide emissions due to product transportation are taken to be those of petroleum diesel, presented in the preceding section.

Product Combustion. Emissions from product combustion are 4,242 tons of carbon dioxide per day.

Net Greenhouse-Gas Emissions. For this example, net greenhouse-gas emissions are negative: For every gallon of diesel fuel produced and burned from a CBTL facility that includes carbon capture and sequestration, between four and seven pounds of carbon dioxide would be removed from the atmosphere. The lower estimate assumes no carbon credit for exported power, while the higher estimate assumes that exported power would displace power generated by existing coal-fired power plants. The key factors that drive this result are the assumption of a 50/50 energy share for coal and biomass and the assumption that 90 percent of net plant-site emissions of carbon dioxide would be captured.

A Model of the Global Liquid-Fuel Market

This appendix documents the policy analysis that underlies the discussion in Chapter Five of how future world petroleum prices might respond to increases in production of alternative liquid fuels. It starts with a brief discussion of factors that shaped our approach to modeling the global market for liquid fuels. It then presents the model and the formulas for OPEC export revenues, consumer surplus, and producer surplus that emerge from that model. It reports numerical findings about how increases in the production of alternative fuels could affect world petroleum prices, OPEC export revenues, and consumer and producer surpluses in the United States. It concludes with a brief discussion of the findings and their policy relevance.

Motivation and Approach

This section explains the basic factors that shaped our approach to modeling the world liquid-fuel market.

We use linear relationships between fuel quantities (supplies and demands) and fuel prices to keep the model simple and transparent. Actual behavioral functions are likely to depart from a linear formulation, but we could not find any empirically based analysis that would predict what form such a departure might take.[1] The simplicity of the model allows a transparency that we prefer, because it allows interested readers to test easily their own beliefs about relevant parameter values and check how sensitive our results are to their beliefs.

A simple behavioral model demands a simple model of OPEC decisionmaking as well. No single model is available to explain all historical OPEC behavior.[2] As a result, we seek to bracket OPEC's behavior so that, even if we cannot explain why OPEC

---

[1] An exploratory analysis of a log-log version of the model indicated that, for the size of the changes in global production we consider, differences in the sizes of price changes predicted by linear and log-log behavioral functions are small. Results from that exploratory analysis are reviewed at the end of this appendix.

[2] For recent econometric tests of alternative models of OPEC behavior, see Alhajji and Huettner (2000a, 2000b), Gulen (1996), Kaufmann et al. (2004), Smith (2005), and Wirl and Kukundzic (2004).

responds in a particular way, we can suggest that the response is likely to fall in a particular range. An extreme assumption might suggest that OPEC will act as a cohesive cartel.[3] More realistic is a model that presents OPEC as an organization with a small core that may shift through time and, from time to time, achieve cohesive action.[4] Still other models suggest that, for the purpose of setting prices, OPEC is simply window dressing and has no long-term effect on prices.[5] Models with noneconomic motivations typically involve competing and shifting goals or goals that lead to complex outcomes with multiple equilibria.[6]

We take an approach that should bracket the various models that might be reasonable without becoming too complex. At one extreme, we assume that a group of OPEC members (the OPEC core) will react cohesively to a long-term increase in unconventional-fuel production. OPEC members outside this core behave as price takers. At the other extreme, we assume that the OPEC core cannot or does not react at all, with the result that the core members as a group maintain the same production levels as they would have in the absence of an increase in unconventional-fuel production. This would occur if reduced excess demand for OPEC exports would require a cohesive reduction in production across the core that the core group could not achieve. Rather, either to maintain budgets or to exploit the difference between the world price and their reservation price during a period of instability, parts of the core break rank in response to lower demand and increase production. In this case, we assume that the portion of the core that remains cohesive is able or willing to cut production only to the extent that net production from the entire core remains constant.

As the analysis that follows shows, world petroleum price falls much more if the OPEC core breaks rank in response to reduced demand for its exports than if it remains cohesive. And the response if the OPEC core is simply a price taker and pure price competitor falls between the two. This is also the case if OPEC members seek to maintain their market share of the total petroleum supply over time, as posited by Gately (2007). So we believe that this range reasonably brackets the most likely responses of OPEC to reduced demand. Our model can accept any specific definition of a core. In the analysis that follows, we assume that Saudi Arabia, Kuwait, Qatar, and the United Arab Emirates comprise the OPEC core.

If the OPEC core can act cohesively, a question remains about how the core members define the reservation price for their petroleum reserves. The most straightforward definition would consider a reservation price equal to the cost of extraction and trans-

---

[3]   Pindyck (1978) posited such a model. It has not held up well over time.

[4]   Adelman (1982) posited such a model. Alhajji and Huettner (2000a, 2000b) offered some empirical evidence for it.

[5]   See, for example, Crémer and Salehi-Isfahani (1989).

[6]   See, for example, Crémer and Salehi-Isfahani (1989) and Teece (1982).

portation to a port of sale.[7] We use a higher price to reflect the fact that OPEC often appears to pursue a conservation-like policy designed to retain petroleum for future generations rather than to maximize any one year's net income.[8]

Our choice of reservation price has no direct effect on our policy-relevant findings. In the analysis that follows, we establish that reservation price by forcing our model to agree with recent EIA price projections using price elasticities that emerge from recent empirical studies. The higher reservation prices that we calculate allow us to reconcile higher absolute values of price elasticities, within the range of recent empirical estimates, with EIA's price projections. These higher absolute values of elasticities tend to increase the level of the effects we measure. But they lie well within the range of reasonable estimates of elasticity values discussed next.

Our model depends heavily on assumptions about future global petroleum prices and price elasticities. Great uncertainty persists about such prices and elasticities. We use EIA projections from 2007 as benchmarks that offer a range for long-term future petroleum prices ($60 to $100 per barrel for a low-sulfur, light oil, such as West Texas Intermediate), primarily because these projections are widely available and likely to be familiar to many readers (EIA, 2007a). We use a range of elasticity estimates drawn from a number of peer-reviewed empirical studies carried out over the past two decades.[9] There is some difficulty reconciling these elasticity estimates with the EIA price projections. We accept this difficulty in our analysis and note that more complete analysis in the future could address this problem more directly.[10]

Before describing the model, we offer a further perspective on our assumptions about long-term oil prices. World oil prices in early 2008 were well above the range that we use in this analysis. As noted in the text, however, we doubt that such high oil prices can be persistently sustained over the future period relevant to our analysis. Though we acknowledge that crude oil prices may continue to be volatile and that, occasionally, crude oil prices might significantly exceed even the very high levels experienced in 2008, the crude oil prices relevant to our analyses are real prices averaged over a period of at least 35 years of ongoing investment and increasing production.

If the high oil prices of 2008 were to persist, oil users would find ways to reduce their demand, and new supplies of alternative fuels would become available *even with-*

---

[7] See, for example, Garber and Nagin (1981).

[8] Noting the conserving behavior of a number of OPEC members that points to a very low social discount rate, Hnyilicza and Pindyck (1976) take a similar approach. Alternatively, a high reservation price could reflect beliefs about petroleum's future value. Expectation of a rising price could lead a rational actor to maintain a reservation price well above and even unrelated to current extraction and transportation costs.

[9] The ranges we chose for elasticity values are based on estimates reported in Cooper (2003), Gately and Huntington (2002), Greene and Ahmad (2005), and Huntington (1991, 1994).

[10] For a fully integrated approach designed to use internally consistent assumptions about the distant future, see Gately (2007).

out government support. For example, over the longer term, users of vehicles can reduce their use, operate their existing vehicles to make them more fuel efficient, and ultimately move to more fuel-efficient vehicles.[11] Our analysis indicates that significant quantities of CTL could become available at long-term prices below $100 per barrel, so prices above this level will only engender more supply. Other fuel sources would also become available, including oil sands, oil shale, and biomass-based renewable sources.

The Model

Definitions

The analysis uses the following variables:

P = world crude petroleum price, defined as the price of an average barrel imported into the United States

P^0 = baseline level of world crude petroleum price that prevails before a change in U.S. policy induces a production increase

P^M = world crude petroleum price that the OPEC core selects when it is cohesive enough to optimize its net revenue or profit from exports

P^F = world crude petroleum price that occurs when the OPEC core cannot react effectively to external changes in production and so freezes its own production in response

D_W = world demand for liquid fuels beyond the demand within the OPEC core

D_W^0 = baseline level of world demand for liquid fuels that prevails before a change in U.S. policy induces a production increase

D_U = U.S. demand for liquid fuels

D_U^0 = baseline level of U.S. demand for liquid fuels that prevails before a change in U.S. policy produces a production increase

S_1 = supply of liquid fuels from the OPEC fringe, i.e., OPEC members outside the core

S_1^0 = baseline level of supply of liquid fuels from the OPEC fringe that prevails before a change in U.S. policy induces a production increase

S_2 = supply of liquid fuels from the rest of the world

S_2^0 = baseline level of supply of liquid fuels from the rest of the world that prevails before a change in U.S. policy induces a production increase

---

[11] Analysts differ in their beliefs about the importance of such effects and how long they will take. To provide some simple perspective, it is useful to keep in mind that, despite having steadily dropped 40 percent in the past 25 years, the energy intensity of the U.S. economy remains more than 30 percent higher than that for comparable countries, such as Japan, Germany, France, and the United Kingdom, suggesting that considerable room remains for demand reductions. This comparison is based on 2000 U.S. dollars and purchasing-power parity.

S_C^0 = baseline level of supply of petroleum from the OPEC core that is available for export before a change in U.S. policy

S_U = supply of liquid fuels from the United States

S_U^0 = baseline level of supply of liquid fuels from the United States that prevails before a change in U.S. policy induces a production increase

D_E = $D_W - S_1 - S_2$ = excess demand outside the OPEC core for liquid fuels from OPEC core members

S_O^M = level of OPEC core production of petroleum that maximizes its net revenue or profit from exports

C = reservation price of petroleum from OPEC core members.

Linear Implementation

A linear implementation yields the following functional forms:

World demand for liquid fuels beyond the demand within the OPEC core
$$D_W = a_W + b_W P$$

Supply from the OPEC fringe
$$S_1 = a_1 + b_1 P$$

Supply from the rest of the world
$$S_2 = a_2 + b_2 P$$

Supply from the United States
$$S_U = a_{US} + b_{US} P$$

Excess demand outside the OPEC core for liquid fuels from OPEC core members
$$D_E = \left(a_W - a_1 - a_2 \right)$$
$$+ \left(b_W - b_1 - b_2 \right) P$$
$$= a_O + b_O P,$$

where $a_O = a_W - a_1 - a_2$

and $b_O = b_W - b_1 - b_2$

Demand from the United States:
$$D_U = a_{UD} + b_{UD} P$$

We assume that demand within the OPEC core is price-inelastic and therefore outside the scope of this model.

We use the following definitions to populate the parameters in the preceding linear functions:

$b_W = \eta_W D_W^0 / P^0$,

where η_W is the worldwide demand elasticity outside the OPEC core

$$b_{UD} = \eta_U \, D_U^0 / P^0,$$

where η_U is the demand elasticity for the United States

$$a_W = D_W^0 - b_W P^0$$

$$b_1 = \varepsilon_1 \, S_1^0 / P^0,$$

where ε_1 is the supply elasticity for fringe OPEC members

$$a_1 = S_1^0 - b_1 P^0$$

$$b_2 = \varepsilon_2 \, S_2^0 / P^0,$$

where ε_2 is the supply elasticity for the rest of the world

$$a_2 = S_2^0 - b_2 P^0.$$

Two Polar Models of OPEC Behavior

OPEC can react to changes in U.S. policy-induced new production in two ways that should bracket its likely range of responses.

Profit Is Optimized. At one extreme, we assume that the OPEC core is able to successfully maximize its profit from petroleum exports by selecting a production level such that

$$\frac{d}{dD_E} D_E (P - C) = 0 \tag{C.1}$$

or, equivalently,

$$P^M + D_E^M / b_0 - C = 0. \tag{C.2}$$

The OPEC fringe reacts as any other competitor would to any price change that follows the OPEC core's reaction to the production induced by the change in U.S. policy. This is a slight variation on the standard price-theoretic model of a monopoly with a competitive fringe.[12] This is the polar case in which the OPEC core reacts as cohesively as possible to any change in production outside of OPEC. By definition, given a competitive fringe, this is as good as it gets for the OPEC core.

From the linear model,

$$P^M = -a_0 / b_0 + D_E^M / b_0, \tag{C.3}$$

---

[12] It maximizes net revenues from exports instead of all net revenues.

which, combined with Equation C.2, can be used to establish the reservation price:

$$C = -a_0/b_0 + 2D_E^M/b_0,$$

(C.4)

and

$$P^M = (1/2)(C - a_0/b_0).$$

(C.5)

Additional production, a_N, of fuel from unconventional sources located outside of OPEC would increase a_2 and, therefore, a_o by a_N. When the OPEC core is able to optimize, both before and after the additional unconventional fuel is introduced, OPEC reacts to the increased production by cutting its price as follows:

$$\Delta P^M = \frac{a_N}{2b_0}.$$

(C.6)

Production Is Frozen. At the other extreme, the OPEC core freezes its crude oil production when liquid-fuel production outside of OPEC causes additional liquid fuels to enter the world market. Why would the OPEC core ever do anything but optimize? One way a failure to optimize might occur is if some members of the OPEC core stop cooperating in response to the downward pressure on world crude oil prices that would be caused by new production of liquid fuels. Instead of reducing consumption to sustain prices, these noncooperating core members may increase their production to offset lower prices, because, even at lower prices, the prevailing prices would likely exceed their marginal costs of production. Doing so will help them sustain revenues at lower world prices. The members of the core that continue to cooperate might attempt to optimize following this desertion, but the net result will be a smaller cut in total OPEC production in response to the new production from the rest of the world and so a lower new world price than if the OPEC core had continued to cooperate and optimized its joint approach. In effect, the members that stop cooperating with the other core members force the members that continue to cooperate to sustain larger cuts than they would have in an optimal, joint core response. We do not attempt to model who stops cooperating; rather, we simply posit a polar case in which the failure of cohesion in the OPEC core is so severe that it can do nothing more than compensate for any increase in production associated with noncooperating members.

If the OPEC core freezes its production, the monopoly pricing scheme no longer applies. Now,

$$D_W = S_1 + S_2 + S_C^0.$$

(C.7)

Inserting the linear relationships, Equation C.7 becomes

$$a_W + b_W P^F = a_1 + b_1 P^F + a_2 + b_2 P^F + S_C^0, \text{ or}$$

$$P^F = \frac{a_W - a_1 - a_2 - S_C^0}{b_W - b_1 - b_2}. \tag{C.8}$$

If S_2 shifts in response to new production at any price level, we have

$$\Delta P^F = \frac{a_N}{b_W - b_1 - b_2}, \tag{C.9}$$

or, equivalently,

$$\Delta P^M = \frac{a_N}{b_0}. \tag{C.10}$$

Comparing Equations C.6 and C.10, any increase in production from the rest of the world will have twice the effect on world prices if the OPEC core freezes its production in response than if it optimizes its response.

The Range Between. Our model of OPEC behavior predicts that the pricing response of the OPEC core to any change, a_N, in world liquid-fuel production outside of the core lies between ΔP^M and ΔP^F.

A third view of the OPEC core is that it acts as a competitor over the longer term. This third view would predict a world price response to new liquid-fuels production within this range. That is because all of OPEC would then react to a price drop by cutting production; because no portion of it would freeze production, price would fall less than it would for our polar freeze case. On the other hand, when price drops in response to lower demand, an effective monopolist always cuts production less than a pure competitor would. So a competitive version of the OPEC core would be more price responsive than our polar monopolist case.

Effects on OPEC Export Revenues and Consumer and Producer Surplus

We treat total OPEC export production, S_T, as the sum of OPEC core exports, S_O, and all OPEC fringe production, minus petroleum consumption within the OPEC fringe. As before, we leave production consumed within the OPEC core outside our model. This measure of total OPEC production is simply the difference between non-OPEC demand and production from the rest of the world, S_2:

$$S_T = \left(D_W - D_F \right) - S_2,$$ (C.11)

where D_F is the petroleum consumption of the countries in the OPEC fringe. We calculate OPEC revenue (R) as the product of world price and total OPEC production:[13]

$$R = S_T P.$$ (C.12)

The change in revenue induced by an increase in production in the rest of the world is

$$\Delta R = S_T^1 P^1 - S_T^0 P^0,$$ (C.13)

where superscripts indicate quantities before (0) and after (1) the increase in production. Substituting from Equation C.11,

$$
\begin{aligned}
\Delta R &= \left[D_W\left(P^1\right) - D_F\left(P^1\right) - S_2\left(P^1\right) \right] P^1 - \left[D_W\left(P^0\right) - D_F\left(P^0\right) - S_2\left(P^0\right) \right] P^0 \\
&= \left(a_W + b_W P^1 - D_F - a_2 - b_2 P^1 \right) P^1 - \left(a_W + b_W P^0 - D_F - a_2 - b_2 P^0 \right) P^0 \\
&= \left(a_W - D_F - a_2 \right)\left(P^1 - P^0 \right) + \left(b_W - b_2 \right)\left[\left(P^1\right)^2 - \left(P^0\right)^2 \right],
\end{aligned}
$$ (C.14)

where we have made the assumption that petroleum consumption within the OPEC fringe is not dependent on the world price. For the price changes that are of interest in this model, this assumption is consistent with the fact that petroleum products sold within OPEC member nations are heavily subsidized.

In this model, a change in world consumer surplus (ΔCS_W) is the area of a trapezoid that is $\left(P^0 - P^1 \right)$ high, $\left(D_W^0 - D_F \right)$ wide on the top, and $\left(D_W^1 - D_F \right)$ wide on the bottom:

$$
\begin{aligned}
\Delta CS_W &= \left(1/2\right)\left(P^0 - P^1 \right)\left(D_W^0 + D_W^1 - 2D_F \right) \\
&= \left(1/2\right)\left(P^0 - P^1 \right)\left(a_W + b_W P^0 + a_W + b_W P^1 - 2D_F \right) \\
&= \left(P^0 - P^1 \right)\left[a_W - D_F + \left(1/2\right) b_W \left(P^0 + P^1 \right) \right].
\end{aligned}
$$ (C.15)

Likewise, a change in the U.S. consumer surplus (ΔCS_U) is given by

---

[13] Transport costs would consume some portion of the OPEC export revenues as they are defined here, although we believe that such effects are small considering the uncertainties with which we are dealing.

$$\Delta CS_U = \left(P^0 - P^1\right)\left[a_{UD} + \frac{1}{2}b_{UD}\left(P^0 + P^1\right)\right],$$

(C.16)

where

$$b_{UD} = \eta_U\left(D_U^0 / P^0\right),$$

in which

η_U = elasticity of demand for liquid fuels in the United States,

D_U^0 = baseline level of demand in the United States,

and

$$a_{UD} = D_U^0 - b_{UD}P^0.$$

By analogy, a change in U.S. producer surplus (ΔPS_U) is measured by

$$\Delta PS_U = \left(P^0 - P^1\right)\left[a_{US} + \frac{1}{2}b_{US}\left(P^0 + P^1\right)\right],$$

(C.17)

where

$$b_{US} = \varepsilon_U\left(S_U^0 / P^0\right),$$

in which

ε_U = elasticity of supply for liquid fuels in the United States

S_U^0 = baseline level of supply in the United States,

and

$$a_{US} = S_U^0 - b_{US}R^0.$$

Numerical Results Under Selected Assumptions

The world liquid-fuel market model, as presented in the preceding section, can accept a wide range of assumptions. This section explains the basis for one set of assumptions and reports the policy implications of these assumptions.

We benchmark the model using the EIA 2007 reference and high-oil-price cases for 2030 (EIA, 2007a), supplemented with assumptions on the consumption of petroleum in the OPEC core, which we want to exclude from the model. We use data on 2006 consumption in our definition of the OPEC core—Saudi Arabia, Kuwait, Qatar, and the United Arab Emirates—from EIA's *Country Analysis Briefs* and project that consumption level forward to 2030 assuming a real annual growth rate of 2.5 percent for the EIA reference case and 2.0 percent for the EIA high-oil-price case. These rates are compatible with growth rates implied by more aggregated world, Organisation for Economic Co-Operation and Development (OECD), non-OECD, and Middle Eastern levels of consumption projected in the two EIA cases that we use as a starting point.[14]

We include all current members of OPEC, except members of the core and Indonesia, in the OPEC fringe category. Since we believe that Indonesia will shortly become a net petroleum importer, we have assumed that it will not be, at least functionally, a member of OPEC. As a practical matter, because our analysis always treats production from the OPEC fringe and from the rest of world in the same way, whether we include Indonesia in the fringe has no effect on our results except for our calculations of OPEC export revenues.

A review of recent empirical estimates of the price elasticity of demand for liquid fuels suggests the following patterns:

- Demand is systematically more elastic within OECD than outside OECD and more elastic within the Group of Seven than in the total OECD.
- The range worth taking seriously for global demand appears to fall between −0.2 and −0.7. Cooper (2003) placed the Group of Seven around −0.2 to −0.5 and the total OECD a bit lower. Gately and Huntington (2002) saw OECD around −0.6 to −0.7 and non-OECD about −0.2 to −0.3. Based on his repeated assessments of markets over the past decade, Greene (in Greene and Ahmad, 2005) chose −0.4 as the best point estimate.
- Based on these recent estimates, for a global price elasticity of demand, we use a range of −0.3 to −0.6, which matches the range that Bartis et al. (2005) used. For the analysis of the consumer surplus, which focuses only on the United States, a higher range, −0.4 to −0.7, is used for η_U, in accordance with information available for highly developed economies.

Our review also found the following patterns for empirical estimates of the price elasticity of non-OPEC supply of liquid fuels:

- Supply has not been given nearly as much attention in recent literature as demand.
- The survey in Huntington (1991) found estimates of the elasticity for United States production of 0.2 to 0.7 with a central tendency in his sample near 0.5. Huntington (1994) suggested a range from 0.2 to 0.6 for global production. He chose 0.4 as the best point estimate. Based on his own updated assessments,

---

[14] See Appendix F in the *International Energy Outlook 2007* (EIA, 2007b). Note that EIA projections may understate future OPEC demand for petroleum, which might result in an overstatement of exports and export revenues (Gately, 2007). We thank Hillard Huntington for this insight, which he provided during his review of our manuscript.

Greene (Greene and Ahmad, 2005) chose 0.3 as the best point estimate for global production.

- Based on these (less) recent estimates, for a global non-OPEC elasticity, a range of 0.2 to 0.6 looks safe. We see no reason to treat the United States differently.

No similar set of estimates is available for the price elasticity of supply in the OPEC fringe. In our analysis, we always apply the same supply elasticity to the fringe that we do to the rest of the world outside OPEC.

Table C.1 summarizes these assumptions. The four columns correspond to EIA's reference and high-oil-price cases and to the elasticity values that lie within the ranges just described and jointly yield the smallest (RAND low) and largest (RAND high) effects on prices.

Table C.2 presents estimates of effects on price and United States consumer, producer, and net surpluses in the alternative cases defined in Table C.1. The results

Table C.1
Assumptions for Alternative Scenarios Examined

EIA Case	Reference		High Oil Price	
RAND Case	Low	High	Low	High
World total production (millions bpd)	117.70	117.70	103.40	103.40
OPEC core consumption (millions bpd)	5.40	5.40	4.80	4.80
OPEC fringe production (millions bpd)	27.70	27.70	21.70	21.70
OPEC fringe consumption (millions bpd)	6.46	6.46	5.76	5.76
Rest of world production (millions bpd)	61.70	61.70	60.70	60.70
U.S. production (millions bpd)	9.42	9.42	11.55	11.55
U.S. consumption (millions bpd)	27.52	27.52	25.00	25.00
Price (U.S. imported crude average, 2005 dollars per barrel)	51.63	51.63	92.93	92.93
Demand elasticity, world	−0.6	−0.3	−0.6	−0.3
Demand elasticity, United States	−0.7	−0.4	−0.7	−0.4
Supply elasticity, OPEC fringe	0.6	0.2	0.6	0.2
Supply elasticity, rest of world	0.6	0.2	0.6	0.2
Implied OPEC reservation price for petroleum[a] (2005 dollars per barrel)	41.86	28.71	79.08	60.27

[a] The model calculates the value of the reservation price that is compatible with the EIA baseline price and the elasticities shown in each case when OPEC maximizes the value of its net income from exports.

Table C.2
Effects on World Crude Oil Price and Annual U.S. Economic Surpluses of Three Million Barrels per Day of Coal-to-Liquids Production (2005 dollars)

EIA Case	Reference		High Oil Price	
RAND Case	Low	High	Low	High
Price drop if OPEC core optimizes (dollars per barrel)	−0.63 (1.2%)	−1.48 (2.9%)	−1.27 (1.4%)	−3.02 (3.2%)
Price drop if OPEC core freezes (dollars per barrel)	−1.27 (2.4%)	−2.98 (5.8%)	−2.55 (2.7%)	−6.04 (6.5%)
Gain in U.S. consumer surplus if OPEC optimizes (billions of dollars)	6.38	15.04	11.68	27.73
Gain in U.S. consumer surplus if OPEC freezes (billions of dollars)	12.82	30.26	23.46	55.81
Loss in U.S. producer surplus if OPEC optimizes (billions of dollars)	2.17	5.11	5.35	12.69
Loss in U.S. producer surplus if OPEC freezes (billions of dollars)	4.32	10.18	10.65	25.29
Gain in U.S. net surplus if OPEC optimizes (billions of dollars)	4.21	9.93	6.33	15.04
Gain in U.S. net surplus if OPEC freezes (billions of dollars)	8.50	20.08	12.81	30.52
Marginal gain in U.S. consumer surplus if OPEC optimizes (dollars per barrel CTL)	5.86	13.81	10.70	25.47
Marginal gain in U.S. consumer surplus if OPEC freezes (dollars per barrel CTL)	11.81	27.95	21.63	51.63
Marginal loss in U.S. producer surplus if OPEC optimizes (dollars per barrel CTL)	1.97	4.65	4.86	11.55
Marginal loss in U.S. producer surplus if OPEC freezes (dollars per barrel CTL)	3.92	9.24	9.64	22.95
Marginal gain in U.S. net surplus if OPEC optimizes (dollars per barrel CTL)	3.89	9.16	5.84	13.92
Marginal gain in U.S. net surplus if OPEC freezes (dollars per barrel CTL)	7.89	18.71	11.99	28.68

reported in the table point to wide ranges in the size of policy-induced effects. Two observations may be useful to understanding the ranges of these effects.

First, as noted, it is very hard to induce the EIA high-oil-price pattern of production, consumption, and prices with the elasticities we consider. Exploratory analysis suggests that the EIA projections imply much smaller elasticities. And discussions with professionals familiar with the details underlying these projections suggest that EIA may well have used much lower elasticities to induce its projections. Given the range of empirical findings reported in the literature, we conclude that we should not consider such low elasticities in our analysis. We report our model's findings for the EIA high-oil-price projections with some skepticism and suggest that readers should be cautious about accepting them. Reasonable ranges for effects on prices—and so on U.S. consumer and producer surpluses and OPEC revenues—probably fall well short of those reported for the high RAND case with the EIA high-oil-price projections. We reflect this expectation by somewhat discounting our findings for the EIA high-oil-price case when suggesting reasonable ranges of results next.

Second, it is unreasonable to expect that all parameter values would take the extreme values that yield a very high or very low effect in either EIA case. It *is* reasonable to expect that demand and supply would be more elastic in the EIA reference case and less elastic in the EIA high-oil-price case. But stepping back, it is most likely that some mix of high and low elasticities would define patterns of supply and demand in 2030 and yield policy effects on prices, consumption, and production well away from the high levels shown for the EIA high-oil-price case or the low levels shown for the EIA reference case.

In sum, we offer the estimates of policy effects in Table C.2 as a place to *start* thinking about reasonable ranges for these effects. In all likelihood, ranges relevant to policymaking lie well within the range offered in the table and probably closer to the low end than the high end. That said, even if we accept such ranges as reasonable, the effects are large enough for policymakers to think seriously about taking actions that would expand production of alternatives to conventional petroleum.

As reported in Table C.2, our analysis shows that an increase in alternative production of three million bpd in 2030 could reduce world petroleum prices by 1.2 percent to 6.5 percent. Given the caveats just reviewed, a more policy-relevant range might be 2 percent to 5 percent.[15] The linearity of the model allows us to interpolate any price effect induced by a different level of alternative production. So even smaller increases in production could well have significant effects.

The table shows that a global increase in new sources of unconventional liquid fuels of three million bpd could induce price reductions that would increase U.S. consumer surplus about $6 billion to $56 billion per year. A portion of this, something like $2 billion to $25 billion, is a transfer from the profits of U.S. producers of liquid fuels. So on net, the price reduction induced by this new production would benefit the United States something in the range of $4 billion to $31 billion a year. The caveats just reviewed suggest that a more reasonable range might be something like $6 billion to $25 billion per year.

---

[15] Exploratory analysis suggests that we are unlikely to observe a combination of parameter values to yield a price change of less than 2 percent or above 5 percent in the EIA reference case. Since we consider it less likely that, during the next few decades, crude oil prices will average at or above the EIA high-oil-price case, we apply a slightly higher reduction to the upper end of the high-oil-price range shown in Table C.2. The subjective range offered for policy purposes reflects these considerations. The ranges for effects on consumer surplus and OPEC export revenues derive from similar considerations.

If new sources of three million bpd came on line, increasing production by one additional barrel would reduce prices enough to increase U.S. consumer surplus by $6 to $52. It would reduce U.S. producer surplus by $2 to $23, leaving a net benefit to the United States of about $4 to $29. Our caveats suggest that a range of $6 to $23 is more reasonable. This value of the net benefit to the United States can be treated as a measure of the classic import externality.[16] The United States should be willing to spend $6 to $23 per barrel more than market prices for substitutes that reduce the demand for production from core OPEC members and thereby put downward pressure on world prices. Put another way, it would be cost-effective to give production of unconventional alternatives to petroleum a subsidy of between $6 and $23 per barrel. This value is not very sensitive to the number of barrels actually produced, so this range is valid over a wide range of levels of new production.

Table C.3 shows that ten million bpd in new global unconventional-fuel production could cut revenue from OPEC exports from $92 billion to $284 billion per year, a range of 11 to 26 percent.[17] If this new capacity came online, one additional barrel of production would reduce OPEC revenues from export by $23 to $73. The

Table C.3
Effects on OPEC Export Revenues Under Selected Assumptions

EIA Case	Reference		High Oil Price	
RAND Case	Low	High	Low	High
Drop in OPEC export revenue with ten million bpd; OPEC optimizes (billions of dollars)	137.10 (16.5%)	174.88 (21.0%)	231.36 (21.2%)	283.74 (26.0%)
Drop in OPEC export revenue with ten million bpd; OPEC freezes (billions of dollars)	92.48 (11.1%)	177.62 (21.4%)	137.28 (12.6%)	261.64 (24.0%)
Marginal drop in OPEC export revenue at ten million bpd; OPEC optimizes (dollars per barrel)	36.47	45.42	69.45	72.77
Marginal drop in OPEC export revenue at ten million bpd; OPEC freezes (dollars per barrel)	24.82	47.70	36.68	69.97

---

[16] This version of an import premium corresponds to the monopsony premium that others have examined in the past. The range estimated here is compatible with ranges estimated in other recent studies reviewed in Leiby (2007), which is an especially helpful review of the literature on this premium. We thank Hillard Huntington for bringing this review to our attention and, more broadly, for his advice on how to approach the import premium.

[17] We consider a ten million bpd increment as the upper bound of the validity of our model. The model results, however, are very close to linear, so effects on OPEC export revenues for unconventional-fuel additions below ten million bpd are easily determined. For example, effects for a three million bpd increment are determined by simply multiplying each of the cells in Table C.3 by 0.3.

caveats reviewed earlier suggest that a more reasonable range might be something like $35 to $70. This value is similar to an externality, but we could treat it directly as an externality only if the United States were willing to pay $1 to reduce revenues from OPEC exports by $1. Suppose the United States were willing to spend only $0.10 to eliminate $1 of revenues from OPEC exports. Then the externality value of this effect would lie in the range of $3 to $7 per barrel.[18] The total value of the externality associated with cutting demand for OPEC production is the sum of the effect on U.S. consumer surplus and U.S. willingness to pay to cut OPEC export revenues. Because both effects derive from the same underlying price effect, high values of one will be associated with high values of the other, and low with low. The range of total values for the externality might be $9 to $27. These are large values relative to the petroleum prices we use in our analysis, which range from $52 to $93.

Preliminary Exploration of a Log-Log Implementation

We implemented a log-log analog to the model just presented to test how sensitive predictions of price effects induced by new sources of unconventional fuel might be to functional form. Table C.4 summarizes our findings. We determined how a ten million bpd increase in production would affect the price of liquid fuels in our eight basic scenarios. Each row of the table shows the results for a different scenario.

Table C.4
Comparison of Linear and Log-Log Implementations: Effects on Price of a Ten Million Barrel per Day Increase in Production

EIA Oil-Price Case	RAND Case	OPEC Model	Price (dollars)			Difference		
			Base	Linear	Log-Log	Dollars	Price (%)	Change (%)
Reference	Low	Optimal	51.63	49.53	49.40	−0.13	0.25	6.1
		Freeze	51.63	47.41	47.59	0.18	0.39	4.2
	High	Optimal	51.63	46.66	45.84	−0.82	1.75	19.3
		Freeze	51.63	41.69	42.70	1.01	2.45	11.9
High	Low	Optimal	92.93	88.68	88.59	−0.09	0.10	1.8
		Freeze	92.93	84.43	84.83	0.40	0.48	3.8
	High	Optimal	92.93	82.86	82.35	−0.51	0.61	5.1
		Freeze	92.93	72.76	75.04	2.28	3.25	11.3

---

[18] Ten cents per dollar implies a willingness to give up only one-tenth of the value of revenue reductions to achieve it. So the size of the externality is only one-tenth the size of the effect.

Several results stand out. First, the differences tend to be small. Relative to the price predicted in the linear model, no difference exceeds 3.25 percent, most are well below this, and the scenario with a difference of 3.25 percent is an extreme one. Percentage differences are larger when stated relative to the price *change* predicted by the linear model. Three differences exceed 10 percent, but, even here, a much smaller difference is more typical. Third, log-log price changes slightly exceed linear price changes when OPEC acts to optimize its net revenue—that is, when we tend to see smaller price changes. But linear price changes slightly exceed log-log price changes when OPEC freezes its production and price changes increase. A linear model thus yields a broader range of outcomes than a log-log model does. Preferring a linear model offers a conservative way to address uncertainty.

In our analysis, the induced price change displayed in Table C.4 sets everything else in motion. Given the exploratory findings on the relative effects of linear and log-log models, we conclude that it is safe to focus on the simpler linear implementation.

Concluding Comments

The principal findings of this analysis are as follows:

- Despite substantial uncertainty about (1) future world petroleum prices, (2) various elasticities of supply and demand, and (3) what exactly motivates OPEC behavior, it is highly likely that a policy-induced increase in the production of unconventional substitutes for petroleum could significantly reduce world petroleum prices.
- The effects of such a price reduction on the well-being of (1) U.S. consumers of petroleum and (2) members of OPEC are large enough to justify U.S. policies that would provide significantly more than the market price for petroleum to induce increased production of unconventional substitutes for petroleum.

Additional analysis along two different tracks might help refine these findings in policy-relevant ways.

The first track would adjust the model itself. For example, it should be worthwhile to adjust it in ways that would reduce some of the tension that currently exists among EIA's projected world prices, elasticities compatible with historical behavior of petroleum markets, and plausible values of OPEC's reservation price of petroleum. The simplest approach would probably be to use a fully integrated model of future world markets that uses reasonable values of price and income elasticities together with reasonable measures of future income to project future world petroleum prices.[19]

---

[19] As noted, substitution of the model that appears to underlie the findings reported in Gately (2007) for the simple model offered here would be one way to do that.

Stepping away from EIA's price projections in this way would, in all likelihood, yield lower future prices and results that were easier to reconcile with the range of long-run elasticities supported by the current empirical literature. Explicit application of such an approach would increase confidence in the impact of potential energy-policy measures on the world oil market.

Given the size of the effects described here, it would also be useful to explore a wider range of functional forms to understand how susceptible the findings reported here are to departures from linearity.

The second track would examine a broader set of effects associated with policy-induced reductions in world petroleum prices. The most obvious would be effects on (1) U.S. producers of petroleum and petroleum substitutes and (2) consumers and producers of petroleum and its substitutes in countries with which we are closely aligned.

The transfer from producers to consumers within the United States includes many different politically relevant transfers, including those among industries, among regions, and among income classes. Political concerns may require mitigation of some of the effects of these transfers; understanding them better would help policymakers weigh alternative mitigation strategies.

Looking beyond the United States, its friends that are net importers of liquid fuel would benefit in a similar way. Those benefits could justify multinational programs to promote unconventional sources of substitutes for petroleum that provided even higher subsidies that those discussed in Chapters Five and Eight. They could also offer particular points of common interest with large, emerging economies, such as China and India, where current competition for scarce liquid fuels tends to be disruptive.

References

Adelman, Morris A., "OPEC as a Cartel," in James M. Griffin and David J. Teece, eds., *OPEC Behavior and World Oil Prices*, London: Allen and Unwin, 1982.

Aimone, Michael, Assistant Deputy Chief of Staff for Logistics, Installations and Mission Support, Headquarters U.S. Air Force, quoted in Masood Farivar, "Military Seeks Oil Savings, Rising Demand, Supply Risks Spur Conservation Move," *Wall Street Journal*, January 9, 2007, p. A15.

———, personal communication with the authors, October 10, 2008.

Alhajji, A. F., and David Huettner, "The Target Revenue Model and the World Oil Market: Empirical Evidence from 1971 to 1994," *Energy Journal*, Vol. 21, No. 2, 2000a, pp. 121–144.

———, "OPEC and World Crude Oil Markets from 1973 to 1994: Cartel, Oligopoly, or Competitive?" *Energy Journal*, Vol. 21, No. 3, 2000b, pp. 31–60.

Altman, Richard L., "Alternative Fuels in Commercial Aviation: The Need, the Approach, Progress," presentation at the 32nd annual FAA Forecasting Conference, Washington, D.C., March 16, 2007. As of June 23, 2008:
http://www.faa.gov/news/conferences_events/aviation_forecast_2007/agenda_presentation/media/9-%20Rich%20Altman.pdf

ANL—*see* Argonne National Laboratory.

Argonne National Laboratory, "The Greenhouse Gases, Regulated Emissions, and Energy Use in Transportation (GREET) Model," GREET 1.8a, August 30, 2007.

ASTM International, *Standard Specification for Aviation Turbine Fuels*, West Conshohocken, Pa., ASTM D1655-08, 2008.

Averitt, Paul, "Coal Resources," in Martin A. Elliott, ed., *Chemistry of Coal Utilization: Second Supplementary Volume*, New York: Wiley, 1981, pp. 55–89.

Barna, Theodore K., Assistant Deputy Under Secretary of Defense, Advanced Systems and Concepts, "OSD Clean Fuel Initiative and the Air Force," briefing, Purdue University, December 2, 2005. As of July 8, 2008:
http://www.purdue.edu/dp/energy/pdf/TBarna-Dec-2005.pdf

Barringer, Felicity, and Andrew R. Sorkin, "In Big Buyout, Utility to Limit New Coal Plants," *New York Times*, February 25, 2007, p. 1.

Bartis, James T., Tom LaTourrette, Lloyd Dixon, D. J. Peterson, and Gary Cecchine, *Oil Shale Development in the United States: Prospects and Policy Issues*, Santa Monica, Calif.: RAND Corporation, MG-414-NETL, 2005. As of June 23, 2008:
http://www.rand.org/pubs/monographs/MG414/

Bechtel, *Baseline Design/Economics for Advanced Fischer-Tropsch Technology*, report prepared for the Federal Energy Technology Center (now National Energy Technology Laboratory), Pittsburgh, Pa., April 1998.

Billings, Kevin W., Deputy Assistant Secretary of the Air Force for Energy, Environment, Safety and Occupational Health, address before the Air and Space Conference, Air Force Association, Washington, D.C., September 25, 2007.

Boerrigter, Howard, and A. van der Drift, "Large-Scale Production of Fischer-Tropsch Diesel from Biomass: Optimal Gasification and Gas Cleaning Systems," in *Congress on Synthetic Biofuels: Technologies, Potentials, Prospects*, Wolfsburg, Germany, 2004.

Boerrigter, Howard, and Robin W. R. Zwart, "High Efficiency Co-Production of Synthetic Natural Gas (SNG) and Fischer-Tropsch (FT) Transportation Fuels from Biomass," *Energy and Fuels*, Vol. 19, No. 2, 2005, pp. 591–597.

BP, *BP Statistical Review of World Energy 2007*, London, June 2007. As of June 23, 2008:
http://www.bp.com/liveassets/bp_internet/globalbp/globalbp_uk_english/reports_and_publications/statistical_energy_review_2007/STAGING/local_assets/downloads/pdf/statistical_review_of_world_energy_full_report_2007.pdf

Brinkman, Norman, Michael Wang, Trudy Weber, and Thomas Darlington, *Well-to-Wheels Analysis of Advanced Fuel/Vehicle Systems: A North American Study of Energy Use, Greenhouse Gas Emissions, and Criteria Pollutant Emissions*, Argonne, Ill.: Argonne National Laboratory, May 2005. As of June 25, 2008:
http://www.transportation.anl.gov/pdfs/TA/339.pdf

Buchanan, T., R. Schoff, and J. White, *Updated Cost and Performance Estimates for Fossil Fuel Power with CO$_2$ Removal*, Palo Alto, Calif.: Electronic Power Research Institute, interim report, EPRI-1004483, December 2002.

Burke, F. P., S. D. Brandes, D. C. McCoy, R. A. Winschel, D. Gray, and G. Tomlinson, *Summary Report of the DOE Direct Liquefaction Process Development Campaign of the Late Twentieth Century: Topical Report*, Oak Ridge, Tenn.: U.S. Department of Energy, Office of Scientific and Technical Information, July 2001.

Camm, Frank, James T. Bartis, and Charles J. Bushman, *Federal Incentives to Induce Early Experience Producing Unconventional Liquid Fuels*, Santa Monica, Calif.: RAND Corporation, TR-586-AF/NETL, 2008.

Chenoweth, Mary E., *The Civil Reserve Air Fleet: An Example of the Use of Commercial Assets to Expand Military Capabilities During Contingencies*, Santa Monica, Calif.: RAND Corporation, N-2838-AF, 1990. As of June 25, 2008:
http://www.rand.org/pubs/notes/N2838/

Chevron Global Aviation, *Aviation Fuels Technical Review*, Houston, Tex., FTR-3, Fall 2004. As of June 23, 2008:
http://www.chevronglobalaviation.com/docs/aviation_tech_review.pdf

———, *Alternative Jet Fuels: Addendum 1 to Aviation Fuels Technical Review*, Houston, Tex.: FTR-3/A1, 2006. As of June 23, 2008:
http://www.chevronglobalaviation.com/docs/5719_Aviation_Addendum._webpdf.pdf

Clark, Nigel, Mridul Gautam, Donald Lyons, Chris Atkinson, Wenwei Xie, Paul Norton, Keith Vertin, Stephen Goguen, and James Eberhardt, *On-Road Use of Fischer-Tropsch Diesel Blends*, SAE Technical Paper Series 1999-01-2251, Warrendale, Pa.: Society of Automotive Engineers, 1999.

Cohen, Linda R., and Roger G. Noll, *The Technology Pork Barrel*, Washington, D.C.: Brookings Institute, 1991.

Cooper, John C. B., "Price Elasticity of Demand for Crude Oil: Estimates for 23 Countries," *OPEC Review*, Vol. 27, No. 1, March 2003, pp. 1–8.

Council on Foreign Relations, *National Security Consequences of U.S. Oil Dependency*, Independent Task Force report 58, John Deutch and James R. Schlesinger, chairs, New York, 2006.

Crémer, Jacques, and Djavad Salehi-Isfahani, "The Rise and Fall of Oil Prices: A Competitive View," *Annales d'Économie et de Statistique*, Vol. 15–16, July–December 1989, pp. 427–454.

Dalton, Matthew, "Big Coal Tries to Recruit Military to Kindle a Market, Use as Liquid Fuel Is an Aim, but Cost, Pollution Are Issues," *Wall Street Journal*, September 11, 2007, p. 15.

Dimotakis, Paul, Robert Grober, and Nate Lewis, *Reducing DoD Fossil-Fuel Dependence*, Fort Belvoir, Va.: Fort Belvoir Defense Technical Information Center, JSR-06-135, September 2006. As of June 24, 2008:
http://handle.dtic.mil/100.2/ADA457233

Dittrick, Paula, "Special Report: Military Testing Fischer-Tropsch Fuels," *Oil and Gas Journal*, Vol. 105, No. 8, February 26, 2007. As of October 1, 2007:
http://www.ogj.com/articles/save_screen.cfm?ARTICLE_ID=285624

DKRW Advanced Fuels, "DKRW Advanced Fuels Secures ExxonMobil MTG Technology, Industrial Siting Permit Granted," press release, Houston, Tex., December 18, 2007. As of June 23, 2008:
http://www.dkrwenergy.com/_filelib/FileCabinet/Media_Kit/MedBowReleaseFinal.pdf?FileName=MedBowReleaseFinal.pdf

DOE—*see* U.S. Department of Energy.

Dunlap, Col. Anne, Air Force Conduct Air, Space, and Cyber Operations, quoted in "Air Force Leaders Explore Ways to Conserve Fuel," American Forces Information Service, Washington, D.C., January 4, 2007.

DuPont, "DuPont Invests $58 Million to Construct Two Biofuels Facilities," press release, London, June 26, 2007. As of June 24, 2008:
http://www2.dupont.com/EMEA_Media/en_GB/newsreleases/article20070626.html

Eastman Chemical Company, "Eastman Announces Key Roles in 2 Major Gulf Coast Gasification Projects," press release, Kingsport, Tenn., July 27, 2007. As of June 23, 2008:
http://www.eastman.com/company/news_center/news_archive/2007/english/corporate_news/financial_news/070727.htm

Edwards, Tim, Don Minus, William Harrison, Edwin Corporan, Matt De Witt, Steve Zabarnick, and Lori Balster, *Fischer-Tropsch Jet Fuels: Characterization for Advanced Aerospace Applications*, paper presented at the 40th AIAA/ASME/SAE/ASEE Joint Propulsion Conference and Exhibit, Fort Lauderdale, Fla.: American Institute of Aeronautics and Astronautics, AIAA-2004-3885, June 2004.

EERE—*see* Office of Energy Efficiency and Renewable Energy.

EIA—*see* Energy Information Administration.

Elliott, Martin A., and G. Robert Yohe, "The Coal Industry and Coal Research and Development in Perspective," in Martin A. Elliott, ed., *Chemistry of Coal Utilization: Second Supplementary Volume*, New York: John Wiley and Sons, 1981, pp. 1–54.

Energy Information Administration, *Coal Data: A Reference*, Washington, D.C., DOE/EIA-0064(93), 1995.

———, *Annual Energy Outlook 2006 with Projections to 2030*, Washington, D.C., DOE/EIA-0383(2006), February 2006a.

————, *Energy Market Impacts of Alternative Greenhouse Gas Intensity Reduction Goals,* Washington, D.C., SR/OIAF/2006-01, March 2006b.

————, *International Energy Annual 2004,* Washington, D.C., July 31, 2006c.

————, *Annual Coal Report 2005,* Washington, D.C., DOE/EIA-0584, October 2006d.

————, *Annual Energy Outlook 2007 with Projections for 2030,* Washington, D.C., DOE/EIA-0383(2007), February 2007a.

————, *International Energy Outlook 2007,* Washington, D.C., DOE/EIA-0484(2007), May 2007b.

————, *Energy Market and Environmental Impacts of S.280, the Climate Change and Stewardship and Innovation Act of 2007,* Washington, D.C., SR-OIAF/2007-04, August 2007c.

————, *Short-Term Energy Outlook August 2007,* Washington, D.C., August 7, 2007d.

————, *Emissions of Greenhouse Gases Report,* Washington, D.C., DOE/EIA-0573(2006), November 28, 2007e.

————, "EIA-819 Monthly Oxygenate Report," *Petroleum Supply Monthly,* Appendix D, Washington, D.C., February 2008a. As of June 23, 2008:
http://www.eia.doe.gov/pub/oil_gas/petroleum/data_publications/petroleum_supply_monthly/historical/2008/2008_02/pdf/819mhilt.pdf

————, *International Petroleum Monthly, February 2008,* Washington, D.C., February 2008b. As of June 23, 2008:
http://www.eia.doe.gov/emeu/ipsr/IPMbackissues.html

————, *Annual Energy Outlook 2008 with Projections for 2030,* Washington, D.C., DOE/EIA-0383(2008), June 2008c.

————, *Annual Energy Review 2007,* Washington, D.C., DOE/EIA-0384(2007), June 2008d.

Energy Research Centre of the Netherlands, "Phyllis, Database for Biomass and Waste," undated Web page. As of June 26, 2008:
http://www.ecn.nl/phyllis/

EPA—*see* U.S. Environmental Protection Agency.

EPA, U.S. Army Corps of Engineers, et al.—*see* U.S. Environmental Protection Agency, U.S. Army Corps of Engineers, et al.

FAA—*see* Federal Aviation Administration.

Fargione, Joseph, Jason Hill, David Tilman, Stephen Polasky, and Peter Hawthorne, "Land Clearing and the Biofuel Carbon Debt," *Science,* Vol. 319, No. 5867, February 29, 2008, pp. 1235–1238.

Farrell, Alexander E., Richard J. Plevin, Brian T. Turner, Andrew D. Jones, Michael O'Hare, and Daniel M. Kammen, "Ethanol Can Contribute to Energy and Environmental Goals," *Science,* Vol. 311, No. 5760, January 27, 2006, pp. 506–508.

Federal Aviation Administration, "Commercial Aviation Alternative Fuels Initiative: Supporting Solutions for Secure and Sustainable Aviation," fact sheet, Washington, D.C., January 3, 2008. As of June 23, 2008:
http://www.faa.gov/news/fact_sheets/news_story.cfm?newsId=10112

Flores, Romeo M., Gary D. Stricker, and Scott A. Kinney, *Alaska Coal Geology, Resources, and Coalbed Methane Potential,* Denver, Colo.: U.S. Department of the Interior, U.S. Geological Survey, digital data series 77, 2004. As of June 23, 2008:
http://bibpurl.oclc.org/web/12293

Foy, Paul, "Two Companies Going After Utah Tar Sands, but the Stuff Is Inferior to Canada's," *Daily Herald Newspaper*, Provo, Utah, September 22, 2006, p. D4.

Frazier, Ian, "Coal Country," *OnEarth*, Spring 2003. As of June 25, 2008: http://www.nrdc.org/onearth/03spr/coal1.asp

Furman, Jason, Jason E. Bordoff, Manasi Deshpande, and Pascal J. Noel, *An Economic Strategy to Address Climate Change and Promote Energy Security*, Washington, D.C.: Brookings Institution, October 2007. As of June 25, 2008: http://www.brookings.edu/~/media/Files/rc/papers/2007/10climatechange_furman/10_ climatechange_furman.pdf

Garber, Steven, and Daniel Nagin, "Deregulation, Synfuels, and the World Price of Crude Oil," *Resources and Energy*, Vol. 3, No. 3, 1981, pp. 223–246.

Gately, Dermot, "Strategies for OPEC's Pricing and Output Decisions," *Energy Journal*, Vol. 16, No. 3, 1995, pp. 1–38.

———, "What Oil Exports Levels Should We Expect from OPEC?" *Energy Journal*, Vol. 28, No. 2, 2007, pp. 151–173.

Gately, Dermot, and Hillard G. Huntington, "The Asymmetric Effects of Changes in Price and Income on Energy and Oil Demand," *Energy Journal*, Vol. 23, No. 1, 2002, pp. 19–56.

Giddings, J. M., B. R. Parkhurst, C. W. Gehrs, and R. E. Millemanne, "Toxicity of a Coal Liquefaction Product to Aquatic Organisms," *Bulletin of Environmental Contamination and Toxicology*, Vol. 25, No. 1, December 1980, pp. 1–6.

Goldberg, Victor P., *Readings in the Economics of Contract Law*, Cambridge and New York: Cambridge University Press, 1989.

Gorbaty, Martin L., Donald F. McMillen, Ripudaman Malhotra, Burtron H. Davis, Francis P. Burke, Harvey D. Schindler, Richard F. Sullivan, Harry Frumkin, David Gray, Glen Tomlinson, and Bary Wilson, "Review of Direct Liquefaction," in H. D. Schindler, ed., *Coal Liquefaction: A Research and Development Needs Assessment*, Vol. II: *Technical Background*, Washington, D.C.: Office of Energy Research, U.S. Department of Energy, DOE/ER-0400, February 1989, pp. 4-1–4-178.

Gray, David, personal communication with the authors, August 6, 2007.

Gray, David, Salvatore Salerno, and Glen C. Tomlinson, *A Techno-Economic Analysis of a Wyoming Located CTL Plant,* Falls Church, Va.: Mitretek, technical report 2005-08, 2005.

Gray, David, and Charles White, telephone communication with David Ortiz, July 23, 2007.

Greene, David Lloyd, and Sanjana Ahmad, *Costs of U.S. Oil Dependence: 2005 Update*, Oak Ridge, Tenn.: Oak Ridge National Laboratory, ORNL/TM-2005/45, January 2005.

Gregor, Joseph H., and Howard E. Fullerton, *Upgrading Fischer-Tropsch Naphtha*, Des Plaines, Ill.: UOP, 1989. As of June 25, 2008: http://www.fischer-tropsch.org/DOE/DOE_reports/90017621/de90017621_toc.htm

Gulen, Gurcan, "Is OPEC a Cartel? Evidence from Cointegration and Causality Tests," *Energy Journal*, Vol. 17, No. 2, 1996, pp. 43–58.

Harlan, James K., *Starting with Synfuels: Benefits, Costs, and Program Design Assessments*, Cambridge, Mass.: Ballinger Publishing Company, 1982.

Harrison, William E. III, senior adviser, Assured Fuels Initiative, "OSD Assured Fuels Initiative: The Drivers for Alternative Aviation Fuels," presentation to 2006 Transportation Research Board 85th Annual Meeting, Washington, D.C., January 22–25, 2006. As of June 23, 2008: http://www.trbav030.org/pdf2006/265_Harrison.pdf

Heinritz-Adrian, Max, Adrian Brandl, Max Hooper, Xinjin Zhao, and Samuel A. Tabak, "An Alternative Route for Coal to Liquids: Methanol-to-Gasoline (MTG) Technology," presentation by Uhde GmbH and ExxonMobil Research and Engineering Company at the Gasification Technologies Conference, San Francisco, Calif., October 17, 2007.

Heinritz-Adrian, Max, Adrian Brandl, Xinjin Zhao, Samuel Tabak, and He Tian Cai, "Production of Gasoline from Coal or Natural Gas by the Methanol-to-Gasoline Process," presented at the Seventh International Exhibition and Conference on Chemical Engineering and Biotechnology, Beijing, 2007.

Hill, Jason, Erik Nelson, David Tilman, Stephen Polasky, and Douglas Tiffany, "Environmental, Economic, and Energetic Costs and Benefits of Biodiesel and Ethanol Biofuels," *Proceedings of the National Academy of Sciences*, Vol. 103, No. 30, July 25, 2006.

Hnyilicza, Esteban, and Robert S. Pindyck, "Pricing Policies for a Two-Part Exhaustible Resource Cartel: The Case of OPEC," *European Economic Review*, Vol. 8, No. 2, August 1976, pp. 139–154.

Huntington, Hillard G., *Inferred Demand and Supply Elasticities from a Comparison of World Oil Models*, Stanford, Calif.: Energy Modeling Forum, Stanford University, working paper 11.5, January 1991.

———, "Oil Price Forecasting in the 1980s: What Went Wrong?" *Energy Journal*, Vol. 15, No. 2, April 1994, pp. 1–22.

IEA—*see* International Energy Agency.

IEA Coal Industry Advisory Board—*see* International Energy Agency Coal Industry Advisory Board.

Illar, Joseph, "Burdened Cost of Fuel," briefing to the Office of the Secretary of Defense, Office of the Director, Program Analysis and Evaluation, Washington, D.C., August 2, 2006.

Interagency Agricultural Projections Committee, *USDA Agricultural Projections to 2016*, Washington, D.C.: U.S. Department of Agriculture, Office of the Chief Economist, OCE-2007-1, 2007.

Intergovernmental Panel on Climate Change, *IPCC Special Report on Carbon Dioxide Capture and Storage*, Cambridge: Cambridge University Press, 2005.

———, *Climate Change 2007: Impacts, Adaptation, and Vulnerability: Contribution of Working Group II to the Fourth Assessment Report of the Intergovernmental Panel on Climate Change*, Cambridge and New York: Cambridge University Press, 2007a.

———, *Climate Change 2007: Mitigation of Climate Change: Contribution of Working Group III to the Fourth Assessment Report of the Intergovernmental Panel on Climate Change*, Cambridge and New York: Cambridge University Press, 2007b. As of June 23, 2008: http://www.ipcc.ch/ipccreports/ar4-wg3.htm

———, *Climate Change 2007: The Physical Science Basis: Contribution of Working Group I to the Fourth Assessment Report of the Intergovernmental Panel on Climate Change*, Cambridge and New York: Cambridge University Press, 2007c. As of June 23, 2008: http://www.ipcc.ch/ipccreports/ar4-wg1.htm

———, "Summary for Policymakers," *Climate Change 2007: Impacts, Adaptation and Vulnerability*, in Intergovernmental Panel on Climate Change, *Climate Change 2007: Impacts, Adaptation and Vulnerability: Contribution of Working Group II to the Fourth Assessment Report of the Intergovernmental Panel on Climate Change*, Cambridge and New York: Cambridge University Press, 2007d, pp. 7–22.

————, *Climate Change 2007: Synthesis Report*, contribution of Working Groups I, II, and III to the Fourth Assessment Report of the Intergovernmental Panel on Climate Change, R. K. Pachauri and A. Reisinger, eds., core writing team, Geneva, Switzerland, 2008. As of June 24, 2008: http://www.ipcc.ch/ipccreports/ar4-syr.htm

International Energy Agency, *World Energy Outlook 2007*, Paris, 2007.

International Energy Agency Coal Industry Advisory Board, *Coal-to-Liquids: An Alternative Oil Supply? Workshop Report, IEA Coal Industry Advisory Board Workshop*, Paris, November 2, 2006.

IPCC—*see* Intergovernmental Panel on Climate Change.

Irvine, A. R., R. M. Wham, J. F. Fischer, R. Salmon, and W. C. Ulrich, *Liquefaction Technology Assessment, Phase II: Indirect Liquefaction of Coal to Gasoline Using Texaco and Koppers-Totzek Gasifiers*, Oak Ridge, Tenn.: Oak Ridge National Laboratory, ORNL-5782, 1984.

Kaufmann, Robert, Stephane Dees, Pavlos Karadeloglou, and Marcelo Sanchez, "Does OPEC Matter? An Econometric Analysis of Oil Prices," *Energy Journal*, Vol. 25, No. 4, 2004, pp. 67–90.

Kinder Morgan, *2005 Annual Report*, Houston, Tex., c. 2006. As of June 23, 2008: http://www.kindermorgan.com/investor/kmi_2005_annual_report_overview.pdf

Klare, Michael T., "The Deadly Nexus: Oil, Terrorism, and National Security," *Current History*, Vol. 101, No. 659, December 2002, pp. 414–420.

Krugman, Paul R., and Maurice Obstfeld, *International Economics: Theory and Policy*, 6th ed., Princeton, N.J.: Addison-Wesley, 2002.

Kuuskraa, Vello A., "CO2-EOR: An Enabling Bridge for the Oil Transition," presentation before the DOE/EPA Workshop on the Economic and Environmental Implications of Global Energy Transitions, Washington, D.C., April 20–21, 2006.

Leiby, Paul N., *Estimating the Energy Security Benefits of Reduced U.S. Oil Imports*, Oak Ridge, Tenn.: Oak Ridge National Laboratory, ORNL/TM-2007/028, February 2007. As of June 24, 2008: http://www.epa.gov/otaq/renewablefuels/ornl-tm-2007-028.pdf

Li Wenbo, "R&D and Industrialization of Coal Liquefaction Technology in China," presentation at the Third International Conference on Clean Coal Technologies for Our Future, Sardinia, May 15–17, 2007.

Maly, Rudolf R., "Effect of GTL Diesel Fuels on Emissions and Engine Performance," presentation at 10th Diesel Engine Emissions Reduction Conference, Coronado, Calif., August 29–September 2, 2004.

Marano, John J., and Jared P. Ciferno, *Life-Cycle Greenhouse-Gas Emissions Inventory for Fischer-Tropsch Fuels*, New York: American Institute of Chemical Engineers, March 2002. As of June 23, 2008: http://internal.lindahall.org/aiche/aiche02/Sp02.119d.pdf

Massachusetts Institute of Technology, *The Future of Coal: Options for a Carbon-Constrained World*, Cambridge, Mass., 2007. As of June 23, 2008: http://web.mit.edu/coal/

Masten, Scott E., "Contractual Choice," in Boudewijn Bouckaert and Gerrit De Geest, eds., *Encyclopedia of Law and Economics*, Vol. III: *The Regulation of Contracts*, Cheltenham, Edward Elgar, 2000. As of June 25, 2008: http://encyclo.findlaw.com/4100book.pdf

Maugeri, Leonardo, *The Age of Oil: The Mythology, History, and Future of the World's Most Controversial Resource*, Westport, Conn.: Praeger Publishers, 2006.

May, Mike, "Development and Demonstration of Fischer-Tropsch Fueled Heavy-Duty Vehicles with Control Technologies for Reduced Diesel Exhaust Emissions," 9th Diesel Engine Emissions Reduction Conference, Newport, R.I., August 24–28, 2003.

Meisel, S. I., "A New Route to Liquid Fuels from Coal," *Philosophical Transactions of the Royal Society*, Series A: *Mathematical, Physical, and Engineering Sciences*, Vol. 300, No. 1453, March 1981, pp. 157–169.

Merrow, Edward W., *An Analysis of Cost Improvement in Chemical Process Technologies*, Santa Monica, Calif.: RAND Corporation, R-3357-DOE, 1989. As of June 23, 2008:
http://www.rand.org/pubs/reports/R3357/

Merrow, Edward W., Kenneth Phillips, and Christopher W. Myers, *Understanding Cost Growth and Performance Shortfalls in Pioneer Process Plants*, Santa Monica, Calif.: RAND Corporation, R-2569-DOE, 1981. As of June 23, 2008:
http://www.rand.org/pubs/reports/R2569/

Meyer, Richard F., E. D. Attanasi, and Philip A. Freeman, *Heavy Oil and Natural Bitumen Resources in Geological Basins of the World*, Reston, Va.: U.S. Geological Survey, open-file report 2007-1084, 2007. As of June 24, 2008:
http://purl.access.gpo.gov/GPO/LPS87698

Milbrandt, A., *A Geographical Perspective on the Current Biomass Resource Availability in the United States*, Golden, Col.: National Renewable Energy Laboratory, NREL/TP-560-39181, December 2005. As of June 24, 2008:
http://www.nrel.gov/docs/fy06osti/39181.pdf

MIT—*see* Massachusetts Institute of Technology.

Mohrig, Jerry R., Christina Noring Hammond, and Paul F. Schatz, *Techniques in Organic Chemistry*, 2nd ed., New York: W. H. Freeman, 2006.

Montgomery, Dave, "Rising Fuel Costs Hit Military Hard," *Fort Worth Star-Telegram*, July 10, 2007, p. C1.

Myers, Christopher W., Ralph F. Shangraw, Mark R. Devey, and Toshi Hayashi, *Understanding Process Plant Schedule Slippage and Startup Costs*, Santa Monica, Calif.: RAND Corporation, R-3215-PSSP/RC, 1986. As of July 10, 2008:
http://www.rand.org/pubs/reports/R3215/

National Energy Board, *Canada's Oil Sands: Opportunities and Challenges to 2015: An Update*, Calgary, Alta., 2006.

National Energy Technology Laboratory, *Major Environmental Aspects of Gasification-Based Power Generation Technologies*, Pittsburgh, Pa., December 2002.

———, *Carbon Sequestration Technology Roadmap and Program Plan*, Pittsburgh, Pa., April 2007a. As of June 23, 2008:
http://www.netl.doe.gov/technologies/carbon_seq/refshelf/project%20portfolio/2007/2007Roadmap.pdf

———, *Baseline Technical and Economic Assessment of a Commercial Scale Fischer-Tropsch Liquids Facility*, Pittsburgh, Pa., DOE/NETL-2007/1260, April 9, 2007b. As of June 23, 2008:
http://204.154.137.14/energy-analyses/pubs/Baseline%20Technical%20and%20Economic%20Assessment%20of%20a%20Commercial%20S.pdf

————, *Fossil Energy Power Plant Desk Reference*, Pittsburgh, Pa., DOE/NETL-2007/1282, May 2007c. As of June 23, 2008:
http://www.netl.doe.gov/energy-analyses/pubs/Cost%20and%20Performance%20Baseline-012908.pdf

————, *Carbon Sequestration Program Environmental Reference Document*, Pittsburgh, Pa., August 2007d. As of June 23, 2008:
http://www.netl.doe.gov/technologies/carbon_seq/refshelf/nepa/

————, *Increasing Security and Reducing Carbon Emissions of the U.S. Transportation Sector: A Transformational Role for Coal with Biomass*, Pittsburgh, Pa., DOE/NETL-2007/1298, August 24, 2007e. As of June 23, 2008:
http://purl.access.gpo.gov/GPO/LPS89719

————, "Gasification Database," Web page, September 2007f. As of June 23, 2008:
http://www.netl.doe.gov/technologies/coalpower/gasification/database/database.html

————, *Chemical-Looping Process in a Coal-to-Liquids Configuration*, Pittsburgh, Pa., DOE/NETL-2008/1307, December 2007g. As of June 23, 2008:
http://www.netl.doe.gov/energy-analyses/pubs/DOE%20Report%20on%20OSU%20Looping%20final.pdf

National Institute for Occupational Safety and Health, *Occupational Hazard Assessment: Coal Liquefaction*, Cincinnati, Ohio: Robert A. Taft Laboratories, March 1981.

National Petroleum Council, *Coal to Liquids and Gas*, working document of the NPC Global Oil and Gas Study, Washington, D.C., topic paper 18, July 2007. As of June 23, 2008:
http://www.npc.org/Study_Topic_Papers/18-TTG-Coals-to-Liquids.pdf

National Research Council, *Coal: Research and Development to Support National Energy Policy*, Washington, D.C.: National Academies Press, 2007. As of June 23, 2008:
http://www.nap.edu/books/030911022X/html/

NETL—*see* National Energy Technology Laboratory.

Nexant, and Cadmus Group, *Environmental Footprints and Costs of Coal-Based Integrated Gasification Combined Cycle and Pulverized Coal Technologies*, Washington, D.C.: U.S. Environmental Protection Agency, Office of Air and Radiation, EPA-430/R-06/006, July 2006. As of June 25, 2008:
http://www.epa.gov/airmarkets/articles/IGCCreport.pdf

NIOSH—*see* National Institute for Occupational Safety and Health.

Norton, Paul, Keith Vertin, Brent Bailey, Nigel N. Clark, Donald W. Lyons, Stephen Goguen, and James Eberhardt, *Emissions from Trucks Using Fischer-Tropsch Diesel Fuels*, SAE Technical Paper Series 982526, Warrendale, Pa.: Society of Automotive Engineers, 1998.

NRC—*see* National Research Council.

OFE—*see* Office of Fossil Energy.

Office of Energy Efficiency and Renewable Energy, U.S. Department of Energy, "Biomass Feedstock Composition and Property Database," Web page, last updated January 25, 2006.

Office of Fossil Energy, U.S. Department of Energy, *Fossil Energy Technology Assessment*, Task Force on Fossil Energy Technology Assessment, Washington, D.C., draft, November 1980.

————, "Recovering 'Stranded Oil' Can Substantially Add to U.S. Oil Supplies," fact sheet, Washington, D.C.: U.S. Department of Energy, 2005. As of June 23, 2008:
http://www.fossil.energy.gov/programs/oilgas/publications/eor_co2/co2_eor_factsheet.pdf

————, "The Early Days of Coal Research," Web page, last updated January 10, 2006. As of June 23, 2008:
http://www.fossil.energy.gov/aboutus/history/syntheticfuels_history.html

Office of Management and Budget, *Guidelines and Discount Rates for Benefit-Cost Analysis of Federal Programs*, Washington, D.C.: Executive Office of the President, Office of Management and Budget, OMB circular A-94, October 29, 1992.

OMB—*see* Office of Management and Budget.

Perlack, Robert D., Lynn L. Wright, Anthony F. Turhollow, Robin L. Graham, Bryce J. Stokes, and Donald C. Erbach, *Biomass as Feedstock for a Bioenergy and Bioproducts Industry: The Technical Feasibility of a Billion-Ton Annual Supply*, Oak Ridge, Tenn.: Oak Ridge National Laboratory, April 2005. As of June 24, 2008:
http://feedstockreview.ornl.gov/pdf/billion%5Fton%5Fvision.pdf

Peterson, D. J., Tom LaTourrette, and James T. Bartis, *New Forces at Work in Mining: Industry Views of Critical Technologies*, Santa Monica, Calif.: RAND Corporation, MR-1324-OSTP, 2001. As of June 23, 2008:
http://www.rand.org/pubs/monograph_reports/MR1324/

Pindyck, Robert S., "Gains to Producers from the Cartelization of Exhaustible Resources," *Review of Economics and Statistics*, Vol. 60, No. 2, April 1978, pp. 238–251.

Public Law 81-774, Defense Production Act, September 8, 1950.

Public Law 109-58, Energy Tax Policy Act, August 6, 2005.

Public Law 110-140, Energy Independence and Security Act, December 19, 2007.

Ralph M. Parsons Company, *Liquefaction Technology Assessment: Final Report*, Oak Ridge, Tenn.: Oak Ridge National Laboratory, ORNL/Sub-7186-30, 1979.

Ratafia-Brown, Jay, Testimony and Prepared Statement to the Committee on Energy and Natural Resources, U.S. Senate, Washington, D.C., S. Hrg. 110-120, May 24, 2007.

Reed, Michael, National Energy Technology Laboratory, personal communication with the authors, June 2007.

Roets, Piet, "Enabling a Coal to Liquids Industry in the U.S.," presentation at the Second Alternative Aviation Fuels Workshop, Atlanta, Ga., October 23, 2006.

Rosborough, Jim, Dow Chemical Company, interview with the authors, July 26, 2007.

Ross, Michael L., "Oil, Islam, and Women," *American Political Science Review*, Vol. 102, No. 1, February 2008, pp. 107–123.

Rubin, Paul H., *Managing Business Transactions: Controlling the Cost of Coordinating, Communicating, and Decision Making*, New York: Free Press, 1990.

Sasol, *Reaching New Energy Frontiers Through Competitive GTL Technology*, corporate brochure, Johannesburg, June 2006.

————, "Project Update: Oryx Gas-to-Liquids (GTL) Joint Venture," press release, Johannesburg, May 22, 2007a.

————, *Investor Insight*, Johannesburg, July 2007b.

Schaberg, Paul, "The Potential of GTL Diesel to Meet Future Exhaust Emission Limits," presentation at the 12th Diesel Engine-Efficiency and Emissions Research Conference, Detroit, Mich., August 20–24, 2006. As of October 5, 2007:
http://www1.eere.energy.gov/vehiclesandfuels/pdfs/deer_2006/session4/2006_deer_schaberg.pdf

Schnepf, Randy, *Agriculture-Based Renewable Energy Production,* Washington, D.C.: Congressional Research Service, RL32712, U.S. Library of Congress, 2006.

Searchinger, Timothy, Ralph Heimlich, R. A. Houghton, Fengxia Dong, Amani Elobeid, Jacinto Fabiosa, Simla Tokgoz, Dermot Hayes, and Tun-Hsiang Yu, "Use of U.S. Croplands for Biofuels Increases Greenhouse Gases Through Emissions from Land-Use Change," *Science,* Vol. 319, No. 5867, February 29, 2008, pp. 1238–1240.

Shanker, Thom, "Military Plans Tests in Search for an Alternative to Oil-Based Fuel," *New York Times,* May 14, 2006, p. A1.

Smith, James L., "Inscrutable OPEC? Behavioral Tests of the Cartel Hypothesis," *Energy Journal,* Vol. 26, No. 1, 2005, pp. 51–82.

Southern States Energy Board, *American Energy Security: Building a Bridge to Energy Independence and to a Sustainable Energy Future,* Norcross, Ga., July 2006. As of June 23, 2008:
http://www.americanenergysecurity.org/studyrelease.html

SSEB—*see* Southern States Energy Board.

State of California, "California Global Warming Solutions Act of 2006," Assembly Bill 32, September 27, 2006.

Stelter, Stan, *The New Synfuels Energy Pioneers: A History of Dakota Gasification Company and the Great Plan Synfuels Plant,* Bismarck, N.D.: Dakota Gasification Company, 2001.

Stranges, Anthony N., "A History of the Fischer-Tropsch Synthesis in Germany 1926–45," in Burtron H. Davis and Mario L. Ocelli, eds., *Fischer-Tropsch Synthesis, Catalysts, and Catalysis,* Amsterdam and Boston: Elsevier, 2007, pp. 1–28.

Swink, Denise, acting director, Office of Energy Assurance, U.S. Department of Energy, testimony on the reauthorization of the Defense Production Act before the Committee on Banking, Housing and Urban Affairs, U.S. Senate, June 5, 2003. As of June 25, 2008:
http://banking.senate.gov/03_06hrg/060503/swink.pdf

Tabak, Samuel, "ExxonMobil Methanol-to-Gasoline: Commercially Proven Route for Production of Synthetic Gasoline," presentation to the 7th U.S-China Oil and Gas Industry Forum, Hangzhou City, September 11–12, 2006. As of January 5, 2008:
http://www.uschinaogf.org/Forum7/7Topic%2023-%20Samuel%20Tabak-%20ExxonMobil-%20 English.pdf

Task Force on Strategic Unconventional Fuels, U.S. Department of Energy, U.S. Department of Defense, and U.S. Department of the Interior, *America's Strategic Unconventional Fuels,* Vol. III: *Resource and Technology Profiles,* Washington, D.C.: Task Force on Strategic Unconventional Fuels, September 2007. As of June 24, 2008:
http://www.unconventionalfuels.org/publications/reports/Volume_III_ResourceTechProfiles(Final).pdf

Teece, David J., "OPEC Behavior: An Alternative View," in James M. Griffin and David J. Teece, eds., *OPEC Behavior and World Oil Prices,* London: Allen and Unwin, 1982, pp. 64–93.

Tilman, David, Jason Hill, and Clarence Lehman, "Carbon-Negative Biofuels from Low-Input High-Diversity Grassland Biomass," *Science,* Vol. 314, No. 5805, 2006, pp. 1598–1600.

Tyson Foods, "ConocoPhillips and Tyson Foods Announce Strategic Alliance to Produce Next Generation Renewable Diesel Fuel," press release, Houston, Tex., and Springdale, Ark., April 16, 2007a. As of June 24, 2008:
http://www.tyson.com/Corporate/PressRoom/ViewArticle.aspx?id=2691

———, "Tyson Foods and Syntroleum Launch Renewable Fuels Venture; Companies to Build United States' First Commercial Synthetic Fuels Plant," press release, Springdale, Ark., June 25, 2007b. As of June 24, 2008:
http://www.tyson.com/Corporate/PressRoom/ViewArticle.aspx?id=2748

UK Department of Trade and Industry, *Coal Liquefaction: Technology Status Report*, London, October 1999.

U.S. Code, Title 10, Section 2306, Kinds of Contracts.

U.S. Department of Energy, "Bioenergy," undated Web page. As of June 23, 2008:
http://www.doe.gov/energysources/bioenergy.htm

———, "DOE Selects Six Cellulosic Ethanol Plants for up to $385 Million in Federal Funding," press release, Washington, D.C., February 28, 2007a. As of June 23, 2008:
http://www.energy.gov/news/4827.htm

———, *Secure Fuels from Domestic Resources: The Continuing Evolution of America's Oil Shale and Tar Sands Industries: Profiles of Companies Engaged in Domestic Oil Shale and Tar Sands Resource and Technology Development*, Office of Naval Petroleum and Oil Shale Reserves, Washington, D.C., June 2007b. As of June 23, 2008:
http://www.fossil.energy.gov/programs/reserves/npr/Secure_Fuels_from_Domestic_Resources_-_P.pdf

U.S. Environmental Protection Agency, "Mountaintop Removal/Valley Fill, Mid-Atlantic Integrated Assessment," Web page, 2007a.

———, 1970–2006 average annual emissions, all criteria pollutants, National Emissions Inventory Air Pollutant Emissions Trends Data, July 2007b. As of July 7, 2008:
http://www.epa.gov/ttn/chief/trends/

U.S. Environmental Protection Agency, U.S. Army Corps of Engineers, U.S. Fish and Wildlife Service, U.S. Office of Surface Mining Reclamation and Enforcement, and West Virginia Department of Environmental Protection, *Mountaintop Mining/Valley Fills in Appalachia: Final Programmatic Environmental Impact Statement*, Philadelphia, Pa.: U.S. Environmental Protection Agency, EPA Region 3, EPA 9-R-05002, October 2005. As of June 25, 2008:
http://www.epa.gov/region3/mtntop/

U.S. Geological Survey, "National Coal Resource Assessment (NCRA): Summary/Overview," undated Web page. As of June 23, 2008:
http://energy.cr.usgs.gov/coal/coal_assessments/summary.html

———, *Natural Bitumen Resources of the United States*, Reston, Va., National Assessment of Oil and Gas fact sheet 2006-3133, November 2006. As of June 23, 2008:
http://pubs.usgs.gov/fs/2006/3133/

USGS—*see* U.S. Geological Survey.

Utah Heavy Oil Program, Institute for Clean and Secure Energy, University of Utah, *A Technical, Economic, and Legal Assessment of North American Heavy Oil, Oil Sands, and Oil Shale Resources*, Salt Lake City, Utah, September 2007. As of June 24, 2008:
http://www.fossil.energy.gov/programs/oilgas/publications/oilshale/HeavyOilLowRes.pdf

Walsh, P. J., E. L. Etnier, and A. P. Watson, "Health and Safety Implications of Alternative Energy Technologies, III: Fossil Energy," *Journal of Environment Management*, Vol. 5, No. 6, November 1981, pp. 483–494.

Wang, Michael Q., *GREET 1.5, Transportation Fuel-Cycle Model*, Vol. 1: *Methodology, Development, Use, and Results,* Argonne, Ill.: Argonne National Laboratory, ANL/ESD-39, August 1999.

Wender, Irving, and Kamil Klier, "Review of Indirect Liquefaction, DOE/ER-0400" in H. D. Schindler, ed., *Coal Liquefaction: A Research and Development Needs Assessment*, Vol. 2, DOE/ER-0400, Washington, D.C.: Office of Energy Research, U.S. Department of Energy, February 1989, pp. 5-1–5-132.

Wham, R. M., J. F. Fisher, R. C. Forrester III, A. R. Irvine, R. Salmon, S. P. N. Singh, and W. C. Ulrich, *Liquefaction Technology Assessment, Phase I: Indirect Liquefaction of Coal to Methanol and Gasoline Using Available Technology*, Oak Ridge, Tenn.: Oak Ridge National Laboratory, ORNL-5664, 1981.

White, Charles, Noblis, telephone communication with David Ortiz, August 24, 2007.

Wicke, Russell, Tech. Sgt., Air Combat Command Public Affairs, "Rising Fuel Costs Tighten Air Force Belt," *Air Force Link*, September 8, 2006. As of June 25, 2008:
http://www.af.mil/news/story.asp?id=123026679

Williams, Robert H., *Toward Cost-Competitive Synfuels from Coal and Biomass with Near-Zero "Well-to-Wheels" GHG Emissions by Simultaneous Exploitation of Two Carbon Storage Mechanisms*, Washington, D.C.: Alternative Fuels Seminar Series, Center for Strategic and International Studies, December 12, 2006.

Williams, Robert H., Eric D. Larson, and Haiming Jin, "Comparing Climate-Change Mitigating Potentials of Alternative Synthetic Liquid Fuel Technologies Using Biomass and Coal," proceedings of the Fifth Annual Conference on Carbon Capture and Sequestration, Alexandria, Va., May 2006a.

———, "Synthetic Fuels in a World with High Oil and Carbon Prices," Proceedings of the Eighth International Conference on Greenhouse Gas Control Technologies, Trondheim, Norway, June 2006b.

Wilson, Bary W., Richard A. Pelroy, D. Dennis Mahlum, Marvin E. Frazier, and Douglas W. Later, "Comparative Chemical Composition and Biological Activity of Single- and Two-Stage Coal Liquefaction Process Streams," *Fuel*, Vol. 63, No. 1, January 1984, pp. 46–55.

Wirl, Franz, and Azra Kukundzic, "The Impact of OPEC Conference Outcomes on World Oil Prices 1984–2001," *Energy Journal*, Vol. 25, No. 1, 2004, pp. 45–62.

World Energy Council, *Survey of Energy Resources 2004*, Oxford, UK: Elsevier Ltd., 2004. As of June 23, 2008:
http://www.worldenergy.org/publications/324.asp

Woynillowicz, Dan, Chris Severson-Baker, and Marlo Raynolds, *Oil Sands Fever: The Environmental Implications of Canada's Oil Sands Rush*, Drayton Valley, Alta.: Pembina Institute, 2005.

Wu, William R. K., and H. H. Storch, *Hydrogenation of Coal and Tar*, Washington, D.C.: U.S. Department of the Interior, Bureau of Mines, 1968.

Wynne, Michael W., Secretary of the Air Force, address before the Air Force Energy Forum, Arlington, Va., March 8, 2007a.

———, "Energy Update," internal memorandum, Washington, D.C., July 14, 2007b.

———, address before the Air and Space Conference, Air Force Association, Washington, D.C., September 24, 2007c.

Yergin, Daniel, *The Prize: The Epic Quest for Oil, Money, and Power*, New York: Simon and Schuster, 1991.